New Uses of Sulfur

New Uses of Sulfur

James R. West, *Editor*

A symposium sponsored by
the Division of Industrial
and Engineering Chemistry
at the 167th Meeting of
the American Chemical
Society, Los Angeles,
Calif., April 2–3, 1974.

ADVANCES IN CHEMISTRY SERIES **140**

AMERICAN CHEMICAL SOCIETY

WASHINGTON, D. C. 1975

Library of Congress CIP Data

New uses of sulfur.
 (Advances in chemistry series; 140 ISSN 0065-2393)

 Includes bibliographies and index.

 1. Sulphur—Congresses.
 I. West, James Robert, 1914-   II. American Chemical Society. III. American Chemical Society. Division of Industrial and Engineering Chemistry. IV. Series.

QD1.A355 no. 140 [TP245.S9] 540′.8s [661′.07′23]
 75-14440
ISBN 0-8412-0218-4   ADCSAJ 140 1-236

Copyright © 1975

American Chemical Society

All Rights Reserved

PRINTED IN THE UNITED STATES OF AMERICA

Advances in Chemistry Series
Robert F. Gould, *Editor*

Advisory Board

Kenneth B. Bischoff

Edith M. Flanigen

Jesse C. H. Hwa

Phillip C. Kearney

Egon Matijević

Nina I. McClelland

Thomas J. Murphy

John B. Pfeiffer

Joseph V. Rodricks

FOREWORD

ADVANCES IN CHEMISTRY SERIES was founded in 1949 by the American Chemical Society as an outlet for symposia and collections of data in special areas of topical interest that could not be accommodated in the Society's journals. It provides a medium for symposia that would otherwise be fragmented, their papers distributed among several journals or not published at all. Papers are refereed critically according to ACS editorial standards and receive the careful attention and processing characteristic of ACS publications. Papers published in ADVANCES IN CHEMISTRY SERIES are original contributions not published elsewhere in whole or major part and include reports of research as well as reviews since symposia may embrace both types of presentation.

CONTENTS

Preface ix

1. Plasticization of Sulfur 1
   B. R. Currell, A. J. Williams, A. J. Mooney, and B. J. Nash

2. Texture of Sulfur Coatings on Urea 18
   G. H. McClellan and R. M. Scheib

3. Sulfur-Coated Urea from a 1-Ton-Per-Hour Pilot Plant 33
   A. R. Shirley, Jr. and R. S. Meline

4. Sulfur in Coatings and Structural Materials 55
   T. A. Sullivan, W. C. McBee, and D. D. Blue

5. Sulfur in Construction Materials 75
   M. A. Schwartz and T. O. Llewellyn

6. Sulfur in Asphalt Paving Mixes 85
   R. A. Burgess and I. Deme

7. Beneficial Use of Sulfur in Sulfur–Asphalt Pavements 102
   D. Saylak, B. M. Gallaway, and H. Ahmad

8. Sulfur/Asphalt Binders for Road Construction 130
   C. Garrigues and P. Vincent

9. Civil Engineering Applications of Sulfur-Based Materials 154
   B. R. Gamble, J. E. Gillott, I. J. Jordaan, R. E. Loov,
   and M. A. Ward

10. Cold Region Testing of Sulfur Foam and Coatings 167
    John M. Dale and Allen C. Ludwig

11. Polyphenylene Sulfide—A New Item of Commerce 174
    R. Vernon Jones and H. Wayne Hill, Jr.

12. Chemical Investigations of Lithium–Sulfur Cells 186
    J. R. Birk and R. K. Steunenberg

13. Metal Sulfide Electrodes for Secondary Lithium Batteries 203
    Laszlo A. Herédy, San-Cheng Lai, Lowell R. McCoy,
    and Richard C. Saunders

14. Sodium–Sulfur Batteries 216
    Lynn S. Marcoux and Eddie T. Seo

Index 231

PREFACE

At the end of World War I, the Frasch sulfur industry in the United States had grown to the point where it could seek a share of the world sulfur market. Until the advent of World War II, it was the source of most of the world's sulfur and provided a dependable supply of this vital commodity for the sulfur-consuming industries.

Throughout the World War II years, great demands were made on the Frasch industry to supply the war effort with its large productive capacity and substantial reserves of sulfur. After the war ended, Frasch sulfur in Mexico, sour gas sulfur in Canada, France, and the United States, and sulfur from expanded refining of sour crudes challenged successfully the pre-World War II dominance of the U.S. Frasch sulfur producers.

Although the world's demand for sulfur has steadily increased in this century, cycles of supply in excess of consumption followed by demand in excess of productive capacity began in the late 1950's. The periods of oversupply stimulated research for uses other than the traditional ones such as sulfuric acid and wood pulping. Groups were formed to conduct research into new areas as well as into areas once abandoned but warranting review because of technology advances.

More recently, knowledgeable observers have begun to predict that with the proper technology and economics, large quantities of sulfur could become available in the 1980's because of environmental regulations which require the desulfurization of fossil fuels. Demand for sulfur is not sensitive to changes in price, and its consumption is not expected to increase as fast as the supply. Also, "involuntary" sulfur—that which is produced regardless of market conditions—has increased as a proportion of total production. As a result, in the 1980's an oversupply of sulfur could reach major proportions in some areas of the world. This has led various industry, governmental, and university groups to initiate or to increase efforts to develop new uses for sulfur.

To use the sulfur which may be recovered from fossil fuels and to be commercially acceptable, a use for sulfur should ideally satisfy the following criteria:

1. The potential tonnage use, realistically assessed, must be large, at a minimum of several hundred thousand tons/yr.
2. The time and cost needed to develop the use must be reasonable.

3. The economics should be favorable, to attract industry interest and capital.

4. The new use must be ecologically acceptable, *i.e.*, it must not create any pollution problems of its own.

In order to stimulate more ideas on new uses for sulfur, it was most appropriate to hold the symposium entitled "Sulfur Utilization—A Status Report." This volume contains most of the papers presented at this refreshing and rewarding conference.

Much of the success of the Symposium can be attributed to D. R. Muir, A. Carpentier, C. B. Meyer, and J. R. Birk, who assumed the roles of discussion leaders. I want to express my thanks to these men whose stimulating comments provided the atmosphere which led to lively and constructive discussions.

Texasgulf, Inc. JAMES R. WEST
New York, N. Y.
December 1974

Plasticization of Sulfur

B. R. CURRELL, A. J. WILLIAMS, A. J. MOONEY, and B. J. NASH

School of Chemistry, Thames Polytechnic, Wellington Street, London SE18 6PF

> *A method, using differential scanning calorimetry, has been developed to estimate quantitatively orthorhombic and monoclinic sulfur in sulfur materials. Sulfur cooled from the melt at 120°C immediately gives monoclinic sulfur which reverts to orthorhombic sulfur within 20 hr. Limonene, myrcene, alloocimene, dicyclopentadiene, cyclododeca-1,5,9-triene, cycloocta-1,3-diene, styrene, and the polymeric polysulfides, Thiokol LP-31, -32, and -33 each react with excess sulfur at 140°C to give a mixture of polysulfides and unreacted sulfur. In some cases substantial amounts of this unreacted sulfur may be held indefinitely in a metastable condition as monoclinic sulfur or "S_8 liquid." Limonene, myrcene, and dicyclopentadiene are particularly effective in retarding sulfur crystallization.*

Sulfur exists in many allotropic forms which differ in their physical and chemical properties. The principal allotropes are orthorhombic (S_α), monoclinic (S_β), and polymeric sulfur (S_w). S_α and S_β are crystalline materials consisting of S_8 rings. S_w consists of chains of up to 10^6 sulfur atoms. S_α is the only allotrope stable under ordinary conditions of temperature and pressure. The sulfur melt below 159°C (S_λ) consists mainly of S_8 rings while above 159°C (the floor temperature of S_w) the melt (S_μ) consists essentially of an equilibrium mixture of S_8 and polymeric sulfur. On rapid cooling from the melt above 159°C "plastic sulfur" is obtained. This is a mixture of polymeric sulfur and "non-crystalline" S_8, and, as its name implies, it possesses plastic properties. These properties rapidly disappear under ordinary conditions of temperature and pressure, and the material becomes brittle.

Elemental sulfur has been proposed for a range of applications which will be fully discussed in other papers in this book, but the development of many of these applications has been hindered (1) by the propensity of sulfur to revert rapidly to the crystalline S_α form.

Many additives have been proposed to modify elemental sulfur to give materials with "plastic" properties. Nearly all of these additives fall under the heading of polymeric polysulfides or, alternatively, substances which may react with elemental sulfur to give *in situ* formation of polymeric polysulfides. There are numerous references in the recent patent literature and also in the pre-1935 literature reviewed by Ellis (*2*).

The "Thiokol" range of polymeric polysulfides have been extensively used as plasticizers. Rueckel and Duecker (*3*) reported the use of polyethylene tetrasulfide (Thiokol A) in a mixture of sulfur (60 parts), Thiokol A (10 parts), and filler (30 parts) for use as a clay pipe jointing material. The addition of Thiokols to sulfur has also been investigated (*4*) in terms of its effect on abrasion resistance, control of film thickness, and adhesion for application as traffic markings. Optimum results were obtained with a loading of 5–15% Thiokol A. Ludwig (*5*) used dipentene and also Thiokol LP-3 to prepare sulfur fiber reinforced composites for joining blocks in wall construction, sulfur–aggregate concretes, and the impregnation of concrete pipes with sulfur to improve their strength and to reduce water permeability.

A patent (*6*) assigned to the Societe Nationale des Petroles d'Aquitaine covers the use of, for example, a polymer made by the interaction of epichlorohydrin, hydrogen sulfide, and alkali or alkaline earth polysulfide in aqueous solution. This polymer, $HS-[CH_2-CHOH-CH_2-S]_nH$ where $n = 4-24$, is mixed with elemental sulfur, a polyolefin (*e.g.* polybutene), and an olefin (*e.g.* styrene). This type of mixture, after heating at 140–160°C, has been used for traffic striping and is said to be serviceable without flaking after 1 year's use, despite heavy traffic.

Tobolsky and Takahashi (*7, 8*) showed that large concentrations of S_8 can remain dissolved in a liquid condition in other polymers. In many cases these compositions appear completely stable, *i.e.*, there is no tendency for the dissolved sulfur to crystallize out. The best example is crosslinked polyethylene tetrasulfide polymers which can retain 40% of dissolved sulfur in the form of liquid S_8 over long periods of time. The sulfur was shown to be S_8 by quantitatively extracting it with carbon disulfide. It was demonstrated that the specific volume of the dissolved sulfur plotted against temperature fits smoothly with the data of specific volume of molten sulfur *vs.* temperature and finally that the mechanical properties of the cross-linked polymers containing dissolved sulfur are just what would be expected from plasticized, cross-linked, amorphous polymers. Ellis (*9*) reported the use of resins made by the interaction of 2,3-xylenol and sulfur monochloride as sulfur additives. These resins were added to three times their weight of molten sulfur. There was no indication of sulfur crystallization in the resultant material, which also ignited with greater difficulty. These sulfur compositions were reported to be useful for moulding and impregnating fiber board.

Other examples of the use of polymeric polysulfide additives include cross-linked polymers (*10*) of the general formula (R—Sx)$_n$, where R is, for example, a methylene radical; HS—(RS)$_n$—H (*11*) where R is a chalcogen interrupted alkylene; and polythioformaldehydes (*12*) HS—(CH$_2$S)$_n$—H.

Examples of additives which would be expected to react with sulfur to form polymeric polysulfides include elemental arsenic and phosphorus. Reaction occurs (*13*) to give a three dimensional network of polysulfide chains with arsenic or phosphorus atoms at the junction points. Low concentrations of arsenic, for example, give a rubbery material whose glass transition temperature increases linearly with increasing arsenic content. Changes in sulfur properties may be obtained by including olefins in the formulation (*6, 14*). Isobutene, diisobutene, octene, cyclooctadiene, and others have been cited as useful. Other unsaturated compounds include drying oil (glyceride of unsaturated fatty acids), asphalts, and pitches. The addition of styrene or ethylene disulfide to sulfur melts at 3% concentration has been investigated by Dale and Ludwig (*15*). Results for both additives were similar. Tensile strengths of 272 and 266 psi were obtained for compositions prepared by heating at 130°C for 2 hrs, casting into bars, and testing after 7 days at ambient temperature. Samples prepared by heating at 187°C for 2 days had the values 135 and 149 psi respectively. A number of patents cover the use of polythiols, including ethylcyclohexanedithiol (*16*), a diester of dithiophosphoric acid (*17*), and dipentene dimercaptan (*18*).

A patent (*19*) assigned to Phillips Petroleum Co. covers the use of di-, tri-, or tetra-thiols which react with sulfur in the presence of basic catalysts such as amines. Typical materials are prepared by heating, at 135°C, a mixture of up to 20 parts by weight of a polythiol containing 0.1 parts tributylamine with 100 parts of sulfur. The resultant mixture was compression-moulded at 24,000 psi and 100°C. Typical results for tensile strength and Shore D hardness are given in Table I. No mention is made of the possible variation in these properties with storage time.

Determination of Sulfur Allotropy

Detection of Polymeric Sulfur. Laser Raman spectroscopy has been used to detect the presence of polymeric sulfur. The Raman spectra of 6N S$_\alpha$ was found to have bands at 84(m), 156(s), 190(w), 220(w), 441(w), and 475(s) cm^{-1}. Polymeric sulfur was prepared by rapidly quenching a sample of 6N sulfur from 180°C to room temperature and

Table I. Results for Tensile Strength and Shore D Hardness[a]

Polythiol Reactant	Polythiol Parts/ 100 Parts Sulfur by Weight	Tensile Strength (psi)	Shore "D" Hardness
None	—	65	80
1,2-Ethanedithiol	10	420	66
1,2-Ethanedithiol	15	310	55
1,2-Propanedithiol	5	200	68
1,2-Propanedithiol	10	142	62
1,2-Propanedithiol	15	303	58
1,2-Propanedithiol	20	239	49
1,2,3-Propanetrithiol	10	730	90

[a] Obtained by Phillips Petroleum Co. (19).

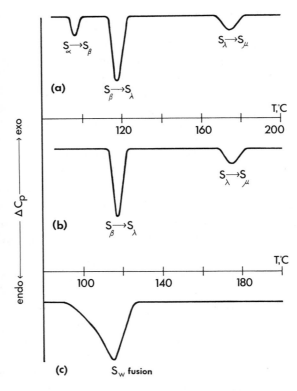

Figure 1. Thermograms obtained by differential scanning calorimetry for a., orthorhombic sulfur (S_α), b., monoclinic sulfur (S_β), c., polymeric sulfur (S_w)

extracting the monomer with carbon disulfide. The spectrum of the polymer has bands at 156(w), 220(w), 278(w), 428(w), and 458(s) cm^{-1}.

Determination of Sulfur Crystallinity. Differential scanning calorimetry has been used to determine the amounts of S_α and S_β present in sulfur materials. The thermogram of S_α (Figure 1a) shows endotherms at 100°C for the transition $S_\alpha \rightarrow S_\beta$, at 119°C for the fusion of S_β ($S_\beta \rightarrow S_\lambda$), and at *ca.* 170°C for the polymerization of S_8 ($S_\lambda \rightarrow S_\mu$). S_β (Figure 1b) shows the endotherms at 119 and *ca.* 170°C. S_λ refers to the sulfur melt below the floor temperature of polymeric sulfur (S_w), and S_μ refers to the equilibrium mixture containing S_w which exists above the floor temperature. The thermogram of S_w (Figure 1c) shows a broad endotherm with a peak at about 118°C, and visual observation showed that the sample was melting to give a viscous product.

The area under the peak representing the transition $S_\alpha \rightarrow S_\beta$ may be used to give the weight of S_α present in the sample. This area is proportional to both the weight of S_α and energy of transition ($\Delta H_g = 0.01254$ Jkg^{-1}). Similarly the S_β fusion peak is proportional to the weight $S_\alpha + S_\beta$ present in the original sample and the energy of fusion ($\Delta H_g = 0.05342$ Jkg^{-1}). Note that any S_α present in the original sample is converted to S_β before fusion.

It is not possible to determine the allotropic nature of sulfur flowers or sulfur which has been rapidly cooled from the melt above 159°C. These samples contain both monomer and polymer, and because of the broad poorly resolved endotherm obtained for monomeric sulfur in the presence of polymer, it is not possible to integrate the area for each fraction, *i.e.* monomer or polymer.

Ten determinations of crystalline sulfur ($S_\alpha + S_\beta$) were carried out on a mixture of sulfur and polymeric polysulfides. An average value of 64.3% with a standard deviation of 2.15 was obtained.

Crystallization of Pure Sulfur

On cooling from the melt at 120°C sulfur crystallizes to S_β at a rate too rapid to be determined by differential scanning calorimetry. On storage at ambient temperature the S_β reverts to S_α (the only allotrope stable under STP conditions). The reversion rate has been measured using differential scanning calorimetry (Figure 2). A 90% reversion rate to S_α was obtained within 10 hr.

As discussed in the previous section, this laboratory has not been able to apply the technique of differential scanning calorimetry to a study of the crystallization of sulfur cooled from the melt above 159°C. Tobolsky [20] has shown that the so-called plastic sulfur, obtained by quick-quenching the sulfur melt from above 159°C, is in a rubbery state

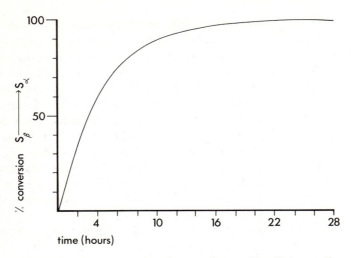

Figure 2. Reversion rate of monoclinic sulfur (S_β) to orthorhombic sulfur (S_α) at ambient temperature

because it contains S_8 rings which are in a metastable liquid state and which act as a plasticizer for the polymeric sulfur, lowering the glass transition temperature to −30°C. Pure polymeric sulfur has a glass transition temperature of 75°C and may be obtained by extracting the quick-quenched sulfur with carbon disulfide. The rapid embrittlement of plastic sulfur is caused by the crystallization of the S_8. This laboratory has shown that in a sample of sulfur quick-quenched from the melt at 180°C, the polymer fraction remains constant at 16.6 ± 0.3% for over 4 wks.

Modified Sulfur Materials

Preparation and Properties. The Thiokols LP-31, -32, and -33 and a range of olefins have been investigated as sulfur additives. Some preliminary results have also been obtained using a 2,3-xylenol/sulfur monochloride resin. Thiokols and olefins were added (25% w/w) to elemental sulfur and the mixture heated at 140°C for 3 hr. In Table II the nature of the final product and the amounts of hydrogen sulfide evolved during heating are given. Of these products only those based on limonene, myrcene, dicyclopentadiene, and cycloocta-1,3-diene appeared to retain any degree of flexibility. Although initially flexible, the Thiokol-based materials became friable in 5–10 days. It was also found that hydrogen sulfide was evolved from these materials at ambient temperature. No hydrogen sulfide evolution was detected at ambient temperature from the olefinic-based products.

Materials based on limonene, myrcene, and styrene were also prepared at 170°C. On addition of the olefin there was a rapid decrease

Table II. Modified Sulfur Materials[a]

Additive	H_2S (mg)	Nature of Product
Thiokol LP-31[b]	1.7	flexible
Thiokol LP-32[b]	12.7	flexible
Thiokol LP-33[b]	22.1	flexible
Alloocimene	1.7	brittle
Cyclododeca-1,5,9-triene	1.9	brittle
Cycloocta-1,3-diene	trace	flexible
Dicyclopentadiene	trace	flexible
Limonene	1.3	flexible
Myrcene	1.4	elastomeric
Octene (1 and 2)	trace	no reaction
Styrene	trace	brittle

[a] Prepared by the addition of olefinic hydrocarbons (5 g) or Thiokols (5 g) to elemental sulfur (20 g) followed by heating for 3 hrs at 140°C.
[b] Structure of "Thiokols" is $HS(C_2H_4OCH_2OC_2H_4S_2)_nC_2H_4OCH_2OC_2H_4SH$. LP-31, $n \cong 49$. LP-32, $n \cong 23$. LP-33, $n \cong 6$.

in the viscosity of the sulfur followed by profuse evolution of hydrogen sulfide (Table III). Qualitative examination indicated that these products were softer and more flexible than the materials prepared at 140°C, even after 9 months.

Tensile strengths have been measured (Table IV) for the limonene, styrene, myrcene, and alloocimene products. In the case of the styrene-based product, a marked increase was obtained between 5 days and 3 months (61–314 psi) with a smaller increase (35–58 psi) for the limonene-based product. These figures are in line with the crystallization rate of the unreacted sulfur in the product (Table V).

Table III. Modified Sulfur Materials[a]

Olefin	H_2S (mg)	Nature of Product
Limonene	277	mastic
Myrcene[b]	—	elastomeric
Styrene[b]	—	mastic

[a] Prepared by the addition of olefinic hydrocarbons (5 g) to elemental sulfur (20 g) followed by heating at 170°C.
[b] Evolution of H_2S was too rapid for determination.

Table IV. Tensile Strength of Modified Sulfur Materials Prepared at 140°C

| Olefin | Tensile Strength (psi) | | |
	5 days	3 months	6 months
Limonene	35	58	123
Styrene	61	314	340
Myrcene	147	—	—
Alloocimene	46	—	—

Table V. Crystallization Rate of Free Sulfur[a]

Percentage of Free Sulphur which has Crystallized ($S_\alpha + S_\beta$)

Additive	2 mos	4 mos	6 mos	8 mos	10 mos	12 mos	14 mos	16 mos	18 mos
Alloocimene	62.0	68.2	72.2	73.5	74.0	75.5	75.5	75.5	75.5
Cyclododeca-1,5,9-triene	75.5	83.0	84.8	85.7	86.5	87.6	88.5	89.7	91.0
Cycloocta-1,3-diene	52.5	60.0	64.0	66.6	68.2	70.8	71.5	73.6	77.6
Dicyclopentadiene	36.3	45.2	49.5	52.7	54.6	55.8	56.3	57.3	57.5
Limonene	31.0	43.0	52.0	55.8	57.6	57.5	58.5	59.6	60.0
Myrcene	45.2	58.5	62.5	64.0	64.7	66.0	66.8	68.3	69.8
Styrene	81.5	85.0	87.6	89.2	90.5	91.6	92.8	93.0	93.0
Thiokol LP-31, Thiokol LP-32, Thiokol LP-33	97% after 2 mos								

[a] 140°C reaction products.

The use of dicyclopentadiene at lower concentrations (5 and 10% w/w) has been examined. These mixtures were heated at 140°C for 3 hr. In both cases only trace amounts of hydrogen sulfide were evolved and brittle products were obtained.

Following the work of Ellis (9) a resin was prepared by the interaction of sulfur monochloride and 2,3-xylenol. This resin (9 g) was added to molten sulfur (27 g), and on cooling a brittle solid was obtained.

Sulfur Crystallization. Each material prepared by the addition of Thiokols or olefins (25 w/w) to sulfur, followed by heating at 140°C, consists of a mixture of polysulfides (40–50%) and unreacted elemental sulfur. Octene is an exception because no reaction occurs and the reactants are recovered. Unreacted elemental sulfur was determined by thin layer chromatography and in each case was shown to be S_8 by its complete extractability with carbon disulfide and laser Raman spectroscopy. Table VI gives the percentage composition of each material after 18 months storage in terms of polysulfides, total unreacted sulfur, and the allotropic composition of the sulfur as determined by differential scanning calorimetry. Figures 3 and 4 show the variation with time of the total crystallinity ($S_\alpha + S_\beta$) of the sulfur. After 18 months all the unreacted sulfur in the Thiokol materials and most of that in the styrene had crystallized, although in these cases there are substantial amounts of the sulfur held as the metastable S_β. Other olefin materials showed much better retardation of sulfur crystallization. After 18 months no S_α has been formed, and although there is some crystallization to S_β, substantial amounts of the sulfur are held in a noncrystalline form ("S_8 liquid"). Table V gives the percentage of the free sulfur that has crystallized (as $S_\alpha + S_\beta$) at various times up to 18 months.

Table VI. Composition of Modified Sulfur Materials After Storage[a]

	Poly-sulfides (%)	Unreacted Free Sulfur (%)			
		S_α	S_β	S_8 (liq)	Total S
D (+) limonene	40.0	0.0	36.0	24.0	60.0
Dicyclopentadiene	45.6	0.0	30.5	23.9	54.4
Alloocimene	39.6	0.0	45.8	14.6	60.4
Myrcene	53.6	0.0	31.2	15.2	46.4
Cycloocta-1,3-diene	47.9	0.0	40.0	12.1	52.1
Cyclododeca-1,5-9-triene	53.7	0.0	38.1	8.2	46.3
Styrene	26.0	60.0	9.0	5.0	74.0
Thiokol LP-31	50.8	17.4	31.8	0.0	49.2
Thiokol LP-32	47.5	24.2	28.3	0.0	52.5
Thiokol LP-33	39.8	36.0	24.2	0.0	60.2

[a] For 18 months at ambient temperature. Note that S_α and S_β are expressed as a percentage of the total composition.

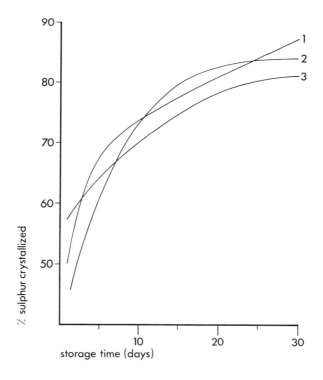

Figure 3. Variation with time of percentage of crystalline sulfur ($S_\alpha + S_\beta$) expressed as a percentage of the total free sulfur in Thiokol/sulfur materials. 1: Thiokol LP-31. 2: Thiokol LP-32. 3: Thiokol LP-33.

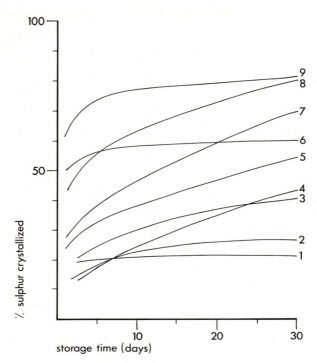

Figure 4. Variation with time of percentage of crystalline sulfur ($S_\alpha + S_\beta$) expressed as a percentage of the total free sulfur in olefin/sulfur materials. 1: limonene. 2: dicyclopentadiene (25% w/w). 3: myrcene. 4: cycloocta-1,3-diene. 5: alloocimene. 6: dicyclopentadiene (5% w/w). 7: cyclododeca-1,3-diene. 8: styrene. 9: dicyclopentadiene (10% w/w).

A 5% w/w loading of dicyclopentadiene has given, after heating at 140°C for 3 hr, a material containing 80% free unreacted sulfur and 20% polymeric polysulfides. A similar 10% loading has given a material containing 79% free unreacted sulfur and 21% polymeric polysulfides. The crystallization rate of this sulfur is given in Figure 4. Unreacted sulfur in the 5% material crystallized less than that in the 10% material. In both cases all the crystalline sulfur is S_β.

In the case of the 170°C products, the percentages of unreacted sulfur were limonene 75.5, myrcene 63.0, and styrene 49.6%. Table VII gives the percentages of this sulfur which crystallized ($S_\alpha + S_\beta$) up to 142 days after preparation.

The material prepared from the sulfur monochloride/2,3-xylenol resin contained 65% unreacted sulfur, 81.5% of which crystallized in 50 days. Heating the resin at 80°C for 1 hr, prior to addition of sulfur

Table VII. Percentage of Free Sulfur Which Crystallized in Materials Prepared at 170°C

	Percentage Sulphur Which Crystallized ($S_\alpha + S_\beta$)			
Additive	3 days	15 days	120 days	142 days
Styrene	—	50.5	68.4	66.0
Limonene	37.4	46.6	58.7	59.5
Myrcene	50.5	47.4	49.5	49.4

gave a product containing 61% unreacted sulfur, 55.8% of which crystallized in 40 days.

Examination of Polysulfide Fractions. Polysulfide fractions from the alloocimene, myrcene, limonene, styrene, and Thiokol LP-33 materials prepared at 140°C have been fractionated by gel permeation chromatography. Other Thiokol products could not be examined because they were insoluble in all organic solvents used. In Tables VIII–XII the molecular weight and molecular formula of each fraction are given. Table

Table VIII. Distribution of Alloocimene Polysulfides

	Fraction				
	A	B	C	D	E
Formula	$(C_{10}H_{15.8}S_{3.2})_x$	$C_{60}H_{96}S_{13}$	$C_{31}H_{54}S_7$	$C_{20}H_{37}S_5$	$C_{20}H_{30}S_2$
MW	2700	1240	656	445	336
Percent of mixture	15.5	11.6	24.3	34.0	14.6

Table IX. Distribution of Myrcene Polysulfides

	Fraction					
	A	B	C	D	E	Insoluble[a]
Formula	$(C_{10}H_{22}S_{8.3})_x$	$C_{53}H_{92}S_{14}$	$C_{30}H_{56}S_{12}$	$C_{20}H_{35}S_6$	$C_{21}H_{41}S_2$	
MW	2700	1200	780	465	351	
Percentage of mixture	26.4	13.8	15.6	24.5	10.0	11.0

[a] Insoluble in CS_2; gave a jelly-like material in CS_2.

Table X. Distribution of Limonene Polysulfides

	Fraction					
	A	B	C	D	E	F
Formula	$C_{112}H_{180}S_{58}$	$C_{79}H_{123}S_{36}$	$C_{60}H_{96}S_{24}$	$C_{40}H_{64}S_{12}$	$C_{20}H_{34}S_6$	$C_{20}H_{35}S_2$
MW	3400	2200	1584	928	464	340
Percentage of mixture	7.3	11.0	12.1	20.6	42.7	6.3

Table XI. Distribution of Styrene Polysulfides

	Fraction		
	A	B	C
Formula	$(C_8H_8)_{35}S_{109}$	$C_{26}H_{28}S_6$	$C_{16}H_{17}S_4$
MW	7100	526	340
Percentage of mixture	74.0	14.9	11.1

Table XII. Distribution of Polysulfides from LP-33/Sulfur Reaction

	Fraction		
	A	B	C
Formula	$C_{205}H_{412}O_{82}S_{122}$		
MW	ca 8000		
Percentage of mixture	71.5	17.8	10.7

Table XIII. Analysis of Principal Polysulfide Fractions

Additive	Allo-ocimene	Myrcene	Limo-nene	Styrene	LP-33
Formula	$C_{20}H_{37}S_5$	$C_{20}H_{34}S_6$	$C_{20}H_{34}S_6$	$(C_8H_8)S_{109}$	$C_{205}H_{412}O_{82}S_{122}$
MW	445	465	464	7100	8000
% Polysulfide sulfur	46.0	30.1	35.3	44.1	43.2
Rank	S_4	S_4	S_4	$S_{3.2}$	$S_{3.05}$
% Thiol content	Nil	Nil	Nil	Nil	1.17
% of Polysulfide mixture	34.0	24.5	42.7	74.0	71.5

XIII gives a more detailed analysis of the principal fractions in terms of sulfur rank and thiol content.

Although alloocimene, myrcene, and limonene products each had small fractions with molecular weights in the order of 3000, in each case the major fraction had a molecular weight in the region of 450.

Weitkamp (21) has reported that the major products from the sulfurization of limonene (mole ratio limonene:sulfur ca. 10:2) were cyclic sulfides of the following structures:

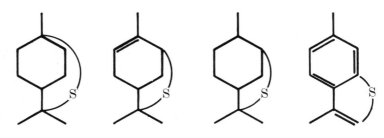

Possible structures for the major limonene product fraction obtained in this laboratory's work may be represented as follows:

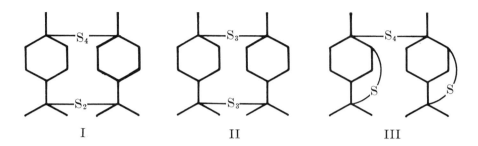

Reduction of I and II would be expected to give a dithiol.

$$\text{I or II} \xrightarrow{\text{red}^n} 2\text{R(SH)}_2 + 2\text{H}_2\text{S}$$

The determined thiol content would then correspond to 66% of the combined sulfur. The value obtained (31.7%), however, suggests a structure similar to III.

$$\text{III} \xrightarrow{\text{red}^n} 2\text{R}^1\text{SH} + 2\text{H}_2\text{S}$$

The silver salt of R¹SH was extracted and the sulfur content found to be 20.5% (calculated for $C_{10}H_{17}S_2Ag$, S = 20.8%). The structure of $C_{20}H_{36}S_6$ is therefore based on III, but the attachment points of the sulfur atoms to the hydrocarbon groups were not determined.

The analysis and sulfur rank of the major styrene product fraction is consistent with a polymer based on the average repeat unit.

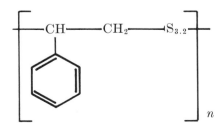

Thiokol LP-33 was the only Thiokol product which could be investigated in detail because all others are insoluble in organic solvents. It gave a polymer with a higher rank and molecular weight compared with Thiokol LP-33. The structure of Thiokol LP-33 is HS—$(C_5H_{10}O_2S_2)_6$—C_2H_4-$OCH_2OC_2H_4SH$; that of the sulfur reaction product is HS—$(C_5H_{10}O_2S_3)_{40}$—$C_2H_4OCH_2OC_2H_4SH$. The thiol content of 1.17% indicates some branching; *i.e.*, no branches would require 0.85% thiol content and one branch 1.23%.

A number of general reaction paths are possible for the reaction of Thiokol polysulfides with sulfur such as:

1. Insertion of sulfur into the sulfur linkages of the polymer

$$HS—(RS_2)_n—RSH + nS_x \rightarrow HS—(RS_{2+x})_n—RSH$$

2. Insertion at the thiol end groups with elimination of hydrogen sulfide

$$HS—(RS_2)_n — RSH + S_x \rightarrow HS—(RS_2)_n —RS_xSH$$
$$HS—(RS_2)_n—RS_x SH + HS—(RS_2)_n—RSH \rightarrow$$
$$HS—(RS_2)_n—RS_x—(RS_2)_n—SH + H_2S$$

Assuming that hydrogen abstraction from the hydrocarbon backbone of the polymer does not occur, the evolved hydrogen sulfide must be formed by interaction of thiol end groups.

Conclusions

On cooling from the melt at 120°C, pure sulfur initially forms metastable S_β which quickly (90% in 10 hr) reverts to the stable S_α. By the use of suitable additives this crystallization may be substantially prevented. Olefins and polymeric polysulfides (Thiokols LP-31, -32, and -33) both react with sulfur to give mixtures of free unreacted sulfur and polysulfides. The lower-molecular weight polysulfides appear to be more effective than those of higher molecular weight in retarding sulfur crystallization. The principal polysulfide fractions formed using alloocimene, myrcene, and limonene have molecular weights of 445, 465, and 464, respectively. In each of these mixtures no S_α is formed, even after 18 mos storage; the unreacted sulfur being a mixture of S_β and "S_8 liquid." Styrene forms a polymeric polysulfide of MW 7100. Here, 60% of the free sulfur has reverted to S_α while Thiokol LP-33 forms a polymeric polysulfide of MW 8000, and 36% of the free sulfur has reverted to S_α. Presumably the controlling factor is the mutual solubility of the elemental sulfur and the polysulfide which depends in part on the molecular weight of the polysulfide. However, although the polysulfides vary in their effectiveness in retarding sulfur crystallization, they all, except in the case of the styrene product, hold substantial proportions of the sulfur as the metastable S_β form.

Experimental

Purification of Sulfur Flowers. Sulfur was purified by the method of Bacon and Fanelli (*22, 23*). Sulfur flowers (1 kg) and magnesium oxide (5 g) were heated to 440°C in a reaction vessel fitted with an air condenser. Boiling was continued under nitrogen for 6 hr to reduce the organic matter in the sulfur to a minimum. The sulfur was then cooled to 130°C and maintained at this temperature for 24 hr. The organic matter settled, and the clear sulfur was filtered off through a glass wool plug. Hydrogen sulfide and sulfur dioxide were removed by heating the filtrate at 120°C/5 mm mercury for 4 hr.

Preparation of Modified Materials. Materials were prepared by heating the Thiokol or olefin (25% w/w) with purified sulfur at 140 or 170°C for 3 hr in a nitrogen atmosphere. In the case of dicyclopentadiene, proportions of 5, 10, and 25% w/w were used. Products were poured into pre-heated (40–50°C) glass dishes and allowed to cool in a desiccator. Liberated hydrogen sulfide from the reaction was trapped in cadmium acetate, and the precipitated cadmium sulfide was determined iodometrically.

2,3-Xylenol/sulfur monochloride resin was prepared by adding sulfur monochloride (26.2 g, 0.194 mole) in toluene (50 ml) to 2,3-xylenol (24.2 g, 0.194 mole) in toluene (250 ml). Hydrogen chloride (13.3 g) was evolved. On removal of solvent at 30°C/15 mm mercury, a yellow resin (36 g) was obtained. The resin (9 g) was added to sulfur (27 g) at 125°C and stirred for 10 minutes until a homogeneous mixture was obtained. The mixture was then poured on to an aluminum foil plate and allowed to cool.

In a further experiment the 2,3-xylenol/sulfur monochloride resin was heated for 1 hr at 80°C. The product (9 g) was added to molten sulfur (27 g) as before.

Determination of Free Sulfur. Free sulfur was obtained by quantitative thin layer chromatography using the method of Davies and Thuraisingham (*24*).

Determination of Combined Sulfur. The sample was decomposed by ignition in an oxygen-filled flask containing 25 ml neutral hydrogen peroxide. The combustion gases were completely absorbed by setting the flask aside for 2 hr. The solution was then boiled to expel carbon dioxide, and the resultant sulfate was determined by alkalimetry.

Determination of Sulfur Rank in Polysulfides. Sulfur rank was determined using the method of Porter, Saville, and Watson (*25*). The sample was reduced with lithium aluminum hydride followed by hydrolysis.

$$RS_xR^1 \ (x \geq 2) \xrightarrow[\text{2. Hydrol.}]{\text{1. LiAlH}_4} RSH + R^1SH + (x-2) H_2S$$

Sulfur rank was then calculated from the results of the determinations of thiols formed and hydrogen sulfide liberated.

Determination of Thiols. Again this depended on the method of Porter, Saville, and Watson (24). Thiol was added to an excess of silver nitrate dissolved in aqueous pyridine. The mixture was then diluted with water, and the pyridinium nitrate formed was titrated with alkali to a phenolphthalein end point.

$$(Py_{12}Ag)^+ + RSH \longrightarrow (n-1)Py + PyH + AgSR$$

Gel Permeation Chromatography. Gel beds used were Bio-Beads S-X2, S-X1, and S-X8. Tetrahydrofuran was used as eluant, and flow rates ranged from 36–90 ml h^{-1} (7.3–18.3 ml cm^{-2} h^{-1}). The sample sizes being collected were *ca.* 2.8 ml.

Molecular Weights. Molecular weights were obtained using an Hitachi–Perkin Elmer 115 or Mechrolab 301A vapor pressure osmometer. Solvents were toluene or chloroform.

Differential Scanning Calorimetry. A Perkin–Elmer DSC-1B was used. Except for the standard (Indium) a scan rate of 4°C/min at a sensitivity of 4 mcal with a chart speed of 20 mm/min was used for the sample. A sensitivity of 8 mcal was used for the standard. Sample weights were 15–20 mg. Static air atmosphere was used.

Acknowledgment

The authors wish to thank the Sulphur Institute for its financial support and also A. Carpentier, H. Fike, and the late R. Leclereq, all of the Sulfur Institute, for their interest and many helpful discussions.

Literature Cited

1. Fike, H. L., ADVAN. CHEM. SER. (1972) **110**, 208.
2. Ellis, C., "The Chemistry of Synthetic Resins," Vol. 2, Reinhold, New York, 1935.
3. Rueckel, W. C., Duecker, W. W., *Bull. Amer. Ceram. Soc.* (1935) **14**, 329.
4. Hancock, C. K., *Ind. Eng. Chem.* (1954) **46**, 2431.
5. Ludwig, A. C., United Nations Report (1969) **TAO/GUA/4**.
6. Societe Nationale des Petroles d'Aquitaine Brit., Patent **1,182,171** (February 25th, 1970).

7. Tobolsky, A. V., Takahashi, N., *J. Polym. Sci. Part A* (1964) **2**, 1987.
8. Tobolsky, A. V., MacKnight, W. J., *Polym. Rev.* (1965) **13**, 1.
9. Ellis, C., U.S. Patents **1,835,766** and **1,835,767** (December 8th, 1931).
10. Dudley, E. A., Grace, N. S., Brit. Patent **1,146,000** (March 19th, 1969).
11. Greco, C. C., Martin, D. J., Can. Patent **831892** (January 13th, 1970).
12. Molinet, G., Audouze, B., Fr. Patent **1,373,025** (November 12th, 1963).
13. Tobolsky, A. V., Owen, G. D. T., Eisenberg, A., *J. Colloid Sci.* (1962) **17**, 717.
14. Dudley, E. A., Grace, N. S., Brit. Patent **1,149,766** (April 23rd, 1969).
15. Dale, J., Ludwig, A. C., *Mater. Res. Std.* (1965) **5**, 411.
16. Louthan, R. P., U.S. Patent **3,434,852** (March 25th, 1969).
17. Signouret, J. B., U.S. Patent **3,459,717** (August 5th, 1969).
18. Stauffer Chemical Co., Brit. Patent **1,200,774** (August 5th, 1970).
19. Phillips Petroleum Co., Brit. Patent **1,149,357** (April 23rd, 1969).
20. Tobolsky, A. V., *J. Polym. Sci. Part C* (1966) (12), 71.
21. Weitkamp, A. W., *J. Amer. Chem. Soc.* (1959) **81**, 3430.
22. Bacon, R. F., Fanelli, R., *Ind. Eng. Chem.* (1942) **34**, 1043.
23. Bacon, R. F., Fanelli, R., *J. Amer. Chem. Soc.* (1943) **65**, 639.
24. Davies, J. R., Thuraisingham, S. T., *J. Chromatogr.* (1968) **35**, 513.
25. Porter, M., Saville, B., Watson, A. A., *J. Chem. Soc.* (1963), 346.

RECEIVED May 1, 1974

2

Texture of Sulfur Coatings on Urea

G. H. McCLELLAN and R. M. SCHEIB

Division of Chemical Development, Tennessee Valley Authority,
Muscle Shoals, Ala. 35660

> *A scanning electron microscope study of pressure- and air-sprayed sulfur coatings, applied to control the release rate of urea as a nitrogen fertilizer, shows that the coatings are mixtures of orthorhombic and polymeric sulfur. The observed sulfur textures vary with the thermal history of the coatings and can be correlated with the performance of the product as a controlled-release nitrogen fertilizer. The satisfactory performance of some sulfur coatings and the unsatisfactory performance of others are related to the presence of desirable sulfur textures and the absence of controllable defects. The principal mechanism of controlled nitrogen release is through pores in the coatings. Sealants can minimize the effects of coating defects.*

A nitrogen fertilizer with controlled-release properties has been the object of much research. Potential advantages of controlled-release nitrogen fertilizers are: increased uptake efficiency by plants; minimized losses by leaching, runoff, or decomposition; reduced application costs from less frequent application; elimination of luxury consumption; prevention of pollution of ground water, streams, and lakes; and prevention of vegetation burning or toxicity damage to seedlings.

Research on controlled-release nitrogen sources is in two broad categories: the synthesis of nitrogen compounds with desired solubility characteristics and the application of protective coatings on soluble nitrogen fertilizers. Examples of synthesized compounds that dissolve slowly are urea–formaldehyde products, isobutylidene diurea, and crotonylidene diurea. However, the relatively high cost of these products tends to limit their consumption to specialty uses.

Attention at the Tennessee Valley Authority (TVA) thus has turned to coating water-soluble nitrogen fertilizers to achieve controlled release. Of the several coating materials tried or considered, sulfur was selected because of cost, application ease, and effectiveness. Urea was selected as the material to be coated as it is the most attractive nitrogen compound because of cost, commercial availability, and high nutrient content. Research at TVA on sulfur coating of urea granules for controlling the dissolution rate began in 1961.

Early work was carried out in the laboratory by coating batches of urea (1). Later the development was extended to a continuous coating process in a small pilot plant with a production rate of about 135 kg/hr (2). At present, the process is being developed further in a large-scale pilot plant with a capacity of about 1 ton/hr, and the study of operating parameters has been quite encouraging.

A flow diagram of the pilot plant as originally built is shown in Figure 1. The process is described in detail in Chapter 3 and consists of

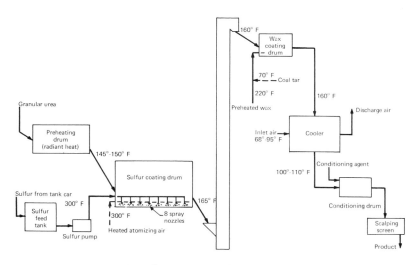

Figure 1. Flow diagram for sulfur-coated urea pilot plant

steps in which preheated urea is sprayed with sulfur in a coating drum. Wax can be applied as a sealant for the coating and conditioners can be added to prevent stickiness in the final product.

Properties of Sulfur Coatings

The textural and solid-state characteristics of sulfur coatings on urea were investigated in the laboratory to identify the controlling factors. The objective was to determine how the interrelated properties of sulfur coatings affect the dissolution mechanism of sulfur-coated urea (SCU) as a controlled-release nitrogen fertilizer. The fine textural details of the sulfur coatings required that the primary investigation be done by scanning electron microscopy (SEM) with supplemental examinations by optical microscopy and x-ray diffraction techniques.

The products described here were produced in the TVA pilot plants and represent a range of dissolution rates that can be correlated with different coating and cooling temperatures, as well as with coating thicknesses. The coatings were applied by air and pressure atomization of the sulfur.

The studies were designed to evaluate the effects of coating uniformity, thickness, and texture, the bonding of the sulfur to the urea surface, the effect of particulate conditioners on the urea, the types of defects in the sulfur coatings, and the function of the wax sealant. The investigation methods that were developed can be applied to products of more advanced sulfur coating technology. The micrographs in this paper were selected from a much larger number of recorded observations to illustrate the textural and structural features that appeared common to representative products.

Nomenclature

There is unequalled confusion in the literature on the nomenclature of sulfur allotropes. Many criteria have been used to name the various forms and preparations of sulfur. As a result, there is no systematic nomenclature. There is not even one allotrope for which a single name is commonly used. Even worse, the same name frequently has been applied to two different forms.

To avoid contributing further to the confusion, the nomenclature suggested by Donohue and Meyer (3) generally will be followed in this work except that when carbon disulfide (CS_2) is used to extract the sulfur coatings, the generic terms polymer, elastomer, and amorphous sulfur will be used interchangeably to describe the insoluble fraction. This nomenclature recognizes that the carbon disulfide-soluble fraction may contain species other than orthorhombic sulfur and that the carbon disulfide-insoluble fraction, known as catenapolysulfur, may contain several sufur species in varying proportions. The notation S_n will be used to represent the carbon disulfide-insoluble material but does not indicate any specific allotropic designation.

Air-Atomized Sulfur Coatings

The exterior surfaces of the air-atomized sulfur coatings characteristically have an uneven, pebbly texture of the type shown in Figure 2. The uneven texture results from the impact and the flow characteristics of the sulfur droplets that accrete to form the coating as the urea particles

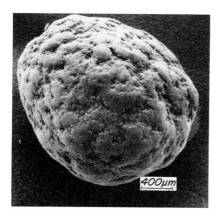
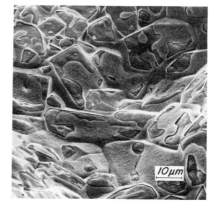

Figure 2. Sulfur coating on air-atomized sulfur-coated urea granule

Figure 3. Mosaic surface texture of sulfur allotropes on the exterior surface

pass through the coating drum. Examination of the surface at higher magnification shows that a mosaic texture develops (Figure 3) when the sulfur allotropes are frozen in place to form a coating that consists of a mixture of cyclo-S_8 rings (othrorhombic sulfur) and S_n polymer chains (elastomer). The detailed view in Figure 4 of a crack cleaving the sulfur coating demonstrates that the intergranular phase of the mosaic texture is distinctly an elastomer form of sulfur. Some of the stretched strands of elastomer sulfur have ruptured where the crack widened beyond their plastic limit. Figure 5 shows a sulfur hull that had been extracted with

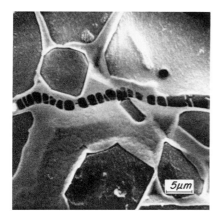
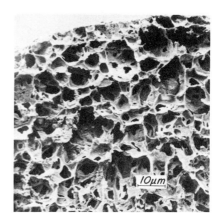

Figure 4. Elastic deformation of amorphous sulfur along a surface crack on the exterior surface

Figure 5. Carbon disulfide-insoluble sulfur elastomer in the cross section of an extracted sulfur coating

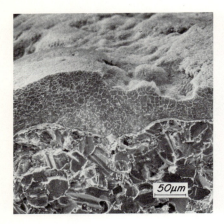 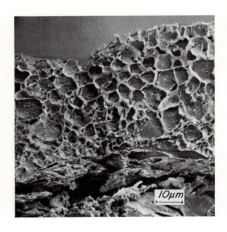

Figure 7.   Cross section of a well developed mosaic texture showing elastic deformation of the intergranular sulfur phase

Figure 6.   Thin spot in a nonuniform sulfur coating

carbon disulfide to remove the carbon disulfide-soluble sulfur, thus exposing the lacy network of insoluble intergranular sulfur elastomer.

Coating uniformity has improved markedly with increased experience in pilot-plant operation. Coating thickness is an important variable in controlling dissolution rate, but this relationship also must consider thickness uniformity. Because thin spots tend to fail easily, the effective coating thickness is only equal to the thinnest areas in the coating. Such thin spots are a process-related defect commonly seen in many nonuniformly coated granules (Figure 6). How much this defect mechanism contributes to high initial dissolution rates depends on the degree of nonuniformity or the amount of undercoated and irregularly coated granules in a particular batch of SCU product. This nonuniformity can be evaluated most effectively by optical microscopic observations.

The structure of the sulfur coatings is best observed in cross section. Cross sections of the coatings are prepared by fracturing granules with a sharp knife edge. The areas studied and reported are not near the pressure points and represent typical observations seen on several granules and not artifacts of specimen preparation.

When SCU products are coated with air-atomized sulfur at about 200°F and quickly cooled, the mosaic texture is well developed, as shown in the cross section of the sulfur coating in Figure 7. The elongation or pulling out of the intergranular sulfur elastomer is clearly shown in Figure 8. When these products are coated at higher temperatures, the product cooling rate has a pronounced effect on the development of sulfur texture. Rapid cooling preserves the elastomer component and

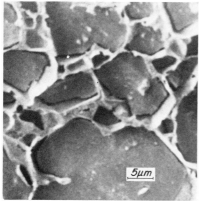
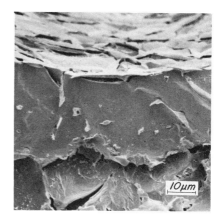

Figure 8. Cross section of a coating showing elongation of sulfur elastomer

Figure 9. Cross section and exterior surface of the massive texture of a slowly cooled sulfur coating applied at high temperature

develops mosaic textures whereas slow cooling from high temperatures promotes the growth of large sulfur crystals, depletes the elastomer, and results in massive textures (Figure 9) that are defective and generally undesirable. Below 190°F the effects of cooling on texture are negligible, presumably because the stable equilibrium phase of orthorhombic sulfur is far more abundant than the metastable elastomer phase, and the conversion of polymeric sulfur to S_8 rings is markedly retarded. Also, the massive textures (Figure 10) that are produced below 190°F do not

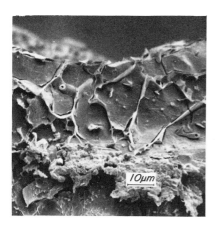

Figure 10. Cross section of massive crystals in a coating produced at a low temperature

Figure 11. Sulfur coating on a pressure-atomized sulfur-coated urea granule

 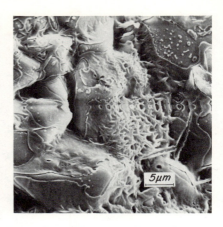

Figure 12. Coated urea granule showing coalescence of sulfur droplets on the exterior surface

Figure 13. Exterior surface of a coating with intergranular pores and poor mosaic texture

develop the tension cracks that have been observed in products coated at higher temperatures (*see* Figure 24).

Pressure-Atomized Sulfur Coatings

The exterior surfaces of the pressure-atomized coatings have an uneven, coarse, pebbly appearance (Figure 11) caused by the aggregation of different-sized sulfur droplets. At higher magnification, the coalescence of the sulfur droplets to form the exterior surface texture is clearly seen (Figure 12), as are some of the larger pores that pit the coating. Surface details show accumulated sulfur crystals along with a poorly developed mosaic texture and numerous intergranular pores (Figure 13). In some areas, the sulfur elastomer extends above the granule surface and is an important local surface feature.

In some samples, a few of the prills were only partially coated. This enabled observation of the contact angle between the sulfur and the conditioned urea surface. The sulfur–surface interfaces in Figure 14 show that the contact angles are positive and less than 90°, indicating that the sulfur is wetting the conditioned urea surface. The sulfur is present as flattened hemi-ellipsoids.

When urea is coated with pressure-atomized sulfur at about 180°F and quickly cooled, a mosaic texture is developed, as in the cross section of the sulfur coating in Figure 15. This mosaic texture is well developed and is characterized by high ratios of the areas of crystalline to carbon disulfide-insoluble sulfur. The mosaic texture in such samples is different

 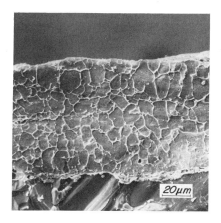

Figure 14. Partially coated urea granule showing positive contact angle between sulfur droplets and the exterior surface of the granule

Figure 15. Cross section of well developed mosaic texture in higher temperature pressure-atomized coating cooled quickly

from that in the air-atomized SCU products previously studied in that the individual sulfur crystallites are poorly defined and the intergranular sulfur phase is present as random sheets rather than as the matrix usually found in SCU coatings (Figures 16 and 7). Detailed study of the surfaces of pressure-atomized sulfur coatings failed to show evidence of plasticity in the intergranular sulfur phases (Figure 17). Thus, the compact texture of such sulfur coatings seems to be a more rigid mosaic of sulfur

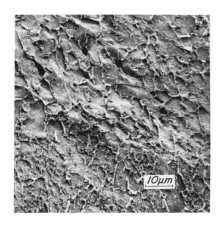 

Figure 16. Cross section showing poorly developed mosaic texture and low elasticity of intergranular sulfur in pressure-atomized sulfur coating

Figure 17. Detail of intergranular sulfur with low elasticity in a cross section

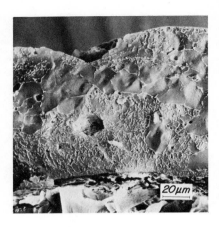

Figure 18. Cross section and exterior (top) surface of massive sulfur texture in high-temperature coating allowed to cool slowly

Figure 19. Layer of sulfur with poorly developed mosaic texture covered with layer of massive sulfur crystals in a cross section

phases, somewhat different from the partially flexible coatings observed in air-atomized sulfur coatings.

When pressure-atomized sulfur coatings are applied at temperatures greater than 180°F and cooled slowly, large sulfur crystals grow at the expense of the polymeric sulfur, resulting in massive sulfur textures that are defective and undesirable (Figure 18). The effect of the cooling rate is negligible in products coated at temperatures below about 180°F. Textures of these products generally are variable mixtures of massive sulfur and poorly developed mosaics. Frequently, the poorly developed mosaic texture occurs along the sulfur–urea interface and may result from quenching of the initial sulfur increments by the cooler urea substrate (Figure 19).

In some pressure-atomized sulfur coatings that have been cooled slowly, a distinct layering of the coating can be seen (Figure 20). Each layer is thought to represent a separate application of a portion of the coating as the product granules pass through the sulfur sprays. The layers are nonuniform because only the side toward the spray is coated. This results in an interlayered or feathered appearance at the edges of the applied increment as shown in A of Figure 20. The sulfur texture in the layers is variable.

Defects in Sulfur Coatings

Sulfur coatings generally show two types of defects. Pinhole capillary pores are the most common defect and appear to be the principal

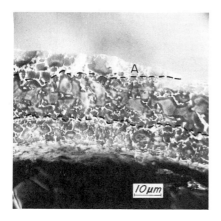 

Figure 20. Cross section showing distinct layers of sulfur with a feathered edge

Figure 21. Pinhole pores in an exterior surface with mosaic texture

path for release of urea from well coated products (Figure 21). Cracks are a common feature only in improperly cooled sulfur coatings and result from large sulfur crystals developing crystallographically controlled strain cracks that penetrate the sulfur coatings and reduce their effective thickness (Figure 22). Rapid quenching tends to freeze the cyclo-S_8:S_n (polymer) distribution, thereby inhibiting crystal growth and retarding any loss of elastomer component. Slow cooling of the products permits an equilibrium shift to the stable phase, crystalline orthorhombic sulfur, which grows at the expense of the metastable sulfur elastomer.

Figure 22. Exterior surface with patterned cracks in an improperly cooled sulfur coating

Figure 23. Spherical void in the cross section of a pressure-atomized sulfur coating

In many of the cross sections of pressure-atomized coatings, spherical voids are common and occasionally have diameters that are a significant fraction of the total coating thickness. This causes a localized decrease in the effective coating thickness (Figure 23) and may reflect gas entrainment associated with the relatively large size of the pressure-atomized sulfur droplets.

In a few cooled products, the pressure-atomized coatings had a series of cracks concentric with the sulfur-urea interface (Figure 24). These cracks obviously are not from applying successive layers of sulfur. They are tension cracks that cut through the sulfur crystals in the coating and result either from strains induced in the coating by the cyclic heating of the particles as they pass through the coating drum or from density changes caused by crystallization of rhombic sulfur. The absence of mosaic texture in the sulfur or the elastomer bond may promote this type of cracking. When there is enough elastomer form of sulfur, the coating should be flexible enough to withstand these thermal or density changes without cracking.

Wax and Conditioner

The wax applied as a sealant on SCU is intended to fill cracks, pinholes, pores, and thin spots in the sulfur coating and thus act in combination with the sulfur to provide a controlled-release coating. The wax sealant primarily fills the surface depressions frequently associated with thin spots in the coating, thereby increasing the effective coating thickness

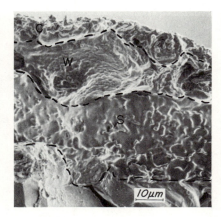

Figure 24. Cross section showing spherical pore and tension cracks concentric with sulfur–urea interface in an improperly cooled sulfur coating

Figure 25. Wax and conditioner filling a thin spot in the sulfur coating. C = conditioner, W = wax, S = sulfur.

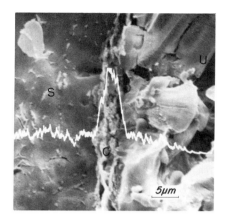 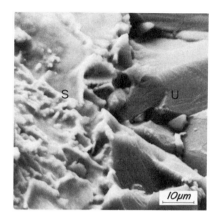

Figure 26. Urea–conditioner–sulfur interface on conditioned urea substrate. U = urea, C = conditioner, S = sulfur.

Figure 27. Urea–sulfur interface on unconditioned urea substrate. U = urea, S = sulfur.

(Figure 25). The distribution of phases in the coating was determined using x-ray area scans of the energy dispersive x-ray system of the SEM.

The urea used in the sulfur-coating process included both prilled and granular forms that usually had been conditioned with clay. The presence of these porous conditioners at the sulfur–urea interface can cause poor bonding of the sulfur to the urea particle, as in Figure 26. The line superimposed on this figure is an electron-probe trace that shows the distribution of aluminum, which marks the location of the clay conditioner across the urea–conditioner–sulfur interface. In products made from unconditioned urea, the sulfur coating is normally tightly bonded (Figure 27), indicating the compatability of these two substances.

Release of Nitrogen from SCU

To observe the release of urea from the SCU products, a number of granules from different samples were soaked in water for 10 min without agitation, carefully filtered, washed with alcohol, air-dried overnight, and examined microscopically for urea release.

As previously mentioned, pinholes in the sulfur coating are the principal path of controlled release. In Figure 28 the ingress of water and subsequent release of urea from such a pinhole are clearly demonstrated. The release of urea along the cracks in the sulfur coating is shown in Figure 29. Large-area failures caused by thin spots also account for the release of urea from nonuniformly coated granules (Figure 30). In Figure 31 a series of swelling cracks is seen in the sulfur coating of a prewetted urea granule. Evidence of such swelling is seen occasionally in

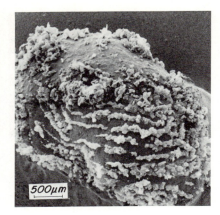

Figure 28. Release of urea through pinhole defects in sulfur coating

Figure 29. Release of urea through cracks in an improperly cooled sulfur coating

soaked granules and is thought to result from internal pressure exerted on the coating by the molar volume increase of the saturated urea solution ($\Delta V = 0.2\%$ at $16°C$) formed when water pentrates the coating and dissolves the urea. Coatings thinner than about 30 μm apparently can rupture under this internal hydrostatic pressure. Short-term laboratory tests have shown that one well developed capillary pore in an otherwise well coated granule is sufficient to vent completely the urea content of that particular granule. The slow release of nitrogen from SCU may result from differences in the times required for individual granules to begin releasing urea through pores. The mechanism of slow release

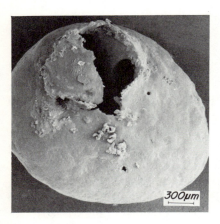
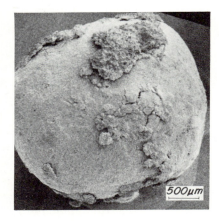

Figure 30. Large area failure caused by thin area in sulfur coating

Figure 31. Swelling cracks in a sulfur coating of a wetted SCU granule

apparently has both diffusion-controlled and nondiffusion-controlled (defect initiated) components operating simultaneously. The gross release rate of a SCU product is an overall averaging of the behavior of many individual granules.

Discussion

The polymeric sulfur allotropes are extremely important in the use of sulfur as a fertilizer coating material, because the allotropes cement the orthorhombic sulfur crystals into mechanically stable coatings. It would be difficult to form a stable coating of wholly crystalline sulfur because of its comparatively poor mechanical properties. Increasing the polymeric sulfur content could improve the coating properties of the sulfur, particularly if the degree of polymerization could be controlled. Too much polymer could make the coating too plastic.

Ludwig and Dale (4) have shown that elemental sulfur, recently melted and allowed to solidify, can be considered a two-phase material consisting of crystalline and polymeric or amorphous sulfur. The crystalline sulfur allotropes may vary from large crystals grown very slowly to small crystals grown very rapidly. Their data show that sulfur samples containing mixtures of crystalline and polymeric sulfur in various proportions have a maximum tensile strength when the mixtures contain 7% polymeric sulfur. Higher or lower polymer contents decrease the strength. Analysis of their findings shows that the polymeric sulfur acts as a typical low modulus of elasticity material that binds the high tensile strength and high modulus of elasticity crystals together. The high strength material constitutes the larger portion of the total mass and is dispersed in a matrix of low modulus materials (5, 6). These observations agree with the textural appearance of sulfur coatings reported here.

The sulfur coatings on a typical pilot-plant product are two-phase composites containing as much as 5% polymeric sulfur. On the basis of the data of Ludwig and Dale, such a coating has about 60% of the maximum tensile strength of a sulfur composite with 7% polymeric sulfur. The polymeric sulfur in the TVA coatings results from sulfur chains that are formed by reactions between the sulfur and impurities in the technical grade sulfur melt.

Summary

The results of this study have shown that the physical appearance of sulfur textures in spray coating can be quite variable and that this variation can be correlated with the performance of the coating in controlling the release of nitrogen from a soluble fertilizer. These examinations also have shown that these sulfur coatings are essentially two-phase composites whose mechanical properties are highly dependent on the thermal history of the sample. The study of these materials has provided information on the mechanism of nitrogen release from the granules as well as the role of conditioners and sealants. Reasons for the satisfactory performances of some coatings and the unsatisfactory performance of others also have been identified.

Because of the small number of samples studied (about 60), as compared with the extremely large number of variables and influencing factors, this can only be considered a preliminary investigation of the textures of sulfur coatings. The potential for the use of sulfur as a coating for controlling the release of soluble fertilizers has been shown elsewhere (*1, 2*). This work has shown some relationships that exist between process variables and sulfur textures. By selecting a combination of conditions, it is possible that the textural properties of sulfur coatings might be used to great advantage in many applications.

Literature Cited

1. Rindt, D. W., Blouin, G. M., Getsinger, J. G., *J. Agr. Food Chem.* (1968) **16**, (5), 773–778.
2. Blouin, G. M., Rindt, D. W., Moore, O. E., *J. Agr. Food Chem.* (1971) **19**, (5), 801–808.
3. Donahue, J., Meyer, B., "The Naming of Sulfur Allotropes," in "Elemental Sulfur" (B. Meyer, Ed.), Interscience, New York, 1965.
4. Ludwig, A. C., Dale, J. M., *Sulfur Inst. J.* (1973) **9**, (1), 10–13.
5. Clauser, H. R., *Sci. Amer.* (July 1973) **229**, (1), 36–44.
6. Slayter, G., *Sci. Amer.* (Jan. 1962) **206**, 124–134.

RECEIVED May 1, 1974

3

Sulfur-Coated Urea from a 1-Ton-Per-Hour Pilot Plant

A. R. SHIRLEY, JR. and R. S. MELINE

Process Engineering Branch, Division of Chemical Development, Tennessee Valley Authority, Muscle Shoals, Ala. 35660

> *Molten sulfur is applied to urea granules by either pneumatic or high-pressure hydraulic atomizing spray nozzles to prepare a controlled-release fertilizer containing 37–40% nitrogen. In a continuous process, urea particles are preheated in a rotary drum. Multiple, thin, concentrically formed layers of sulfur are applied in a second rotary drum. Wax is applied in a third drum to seal minute imperfections in the sulfur coating. The granules are quick-cooled in a fluid bed cooler and lightly dusted with conditioner. Operating conditions and processing temperatures depend on the sulfur atomization process. Some major advantages of hydraulic spraying include up to 98% decrease in sulfur dust and mist in the sulfur-coating drum and doubled plant production capacity. Sulfur-coated urea is being tested in 50 states and 54 countries, showing significant advantages with several crops.*

For several years the Tennessee Valley Authority (TVA) has been encapsulating soluble fertilizer granules and prills with sulfur to impart controlled-release properties (1, 2, 3). Most of the work, both bench and pilot scale, has centered on coating urea, $CO(NH_2)_2$, with sulfur. In the pure state, urea contains 46.67% nitrogen, making it the most concentrated solid nitrogen fertilizer extensively produced. Furthermore, urea can be manufactured economically and is a good source of nitrogen for many plants, grasses, and trees. It is becoming the world's leading nitrogen fertilizer. Our research and development effort has shown it to be a very desirable base for a slow-release fertilizer.

If urea has been properly coated with sulfur, agronomists have found that it possesses significant advantages over more conventional fertilizers when used with certain crops or in certain climatic or soil conditions (4, 5, 6). In field tests with rice, sugar cane, pineapple, and forage crops, sulfur-coated urea showed advantages over uncoated urea or ammonium nitrate. Leaching losses have been significantly reduced in some tests (7). In other tests, luxury nitrogen consumption has been greatly reduced or eliminated, and vegetation burning and damage to seedlings have been prevented. Exceptionally good results have been obtained when sulfur-coated urea was used to fertilize lawns and ornamental shrubs. In almost all cases these results can be obtained at a lower cost because fewer fertilizer applications are required.

Aside from the agronomic advantages, sulfur-coated urea has several physical advantages over conventional nitrogen fertilizers. It resists caking in humid conditions. Uncoated urea must be protected from caking by incorporating formaldehyde or applying a parting agent, such as a kaolinite clay, on the solid granules. Sometimes even these precautions are not effective enough to prevent caking. Sulfur-coated urea is compatible with other fertilizers such as superphosphates and may be used in a wider range of blend formulations than uncoated urea which is not compatible with superphosphate.

Cost estimates have indicated that sulfur-coated urea can be produced for about one-third more per unit of nitrogen than straight urea. This would be much cheaper than other controlled-release products now marketed such as urea formaldehyde or isobutylene diurea (IBDU).

Product Description

Conventional sulfur-coated urea consist of granules or prills of urea uniformly coated with a layer of sulfur. A wax coating is applied over the sulfur coating to fill pores, cracks, or thin spots. For most pilot studies, TVA used a microcrystalline wax containing about 10% oil. A conditioner is applied to the waxed granules to absorb this oil and thus provide good flow properties to the final product. The amounts of sulfur, wax, and conditioner applied to the urea depend on the desired nitrogen dissolution rate. These amounts vary with the processing conditions, the size and shape of the substrate urea, and the type of wax and conditioner used. The typical product which has a nominal size range of minus 6 plus 9 Tyler standard sieve size contains 82% urea, 1% clay conditioner (applied as a parting agent by the urea supplier), 13% sulfur, 2% wax, and 2% diatomaceous earth. In the past, the sulfur-coated urea has also contained 0.25% coal tar applied as a microbicide to reduce the soil microbial attack on the wax. However, recent agronomic tests have demonstrated that the coal tar has very little value. It was therefore discarded in favor of a more simplified process which produced a final product with better handling characteristics.

Recently, TVA has produced a slow-release urea coated only with sulfur (8, 9). This material, which also has a nominal size range of minus 6 plus 9 Tyler standard sieve size, typically contains 80% urea, 1% pre-applied parting agent, and 19% sulfur. There is much less agronomic data on this product than on the conventional product, and its effectiveness as a slow-release nitrogen fertilizer is not fully known, although early reports are very encouraging.

The comparative advantage of controlled-release sufur-coated urea is measured by determining the percentage of urea that will dissolve when a 50-g sample is immersed in 250 g of water for 7 days at 100°F. The quality of the coating is determined by comparing this dissolution rate with the total coating of sulfur, wax, coal tar, and conditioner applied to the urea. Acceptable products dissolve 20–30% in 7 days and 1% or less daily thereafter. The dissolution rate can be varied by changing the coating weight. Soil and climate conditions and the growing season of the crop being fertilized are other variables that might require higher or lower dissolution rates.

As might be expected, the crystalline makeup and the uniformity of the sulfur coating are important, even more so for the sulfur-only material (10). The sulfur structure must be elastic and the urea must be evenly coated. Operating conditions must be controlled to give this structure. McClellan fully discusses the subject of sulfur crystalline structure in Chapter 2.

Raw Materials

Urea. The physical characteristics of the substrate material are very important in producing any coated material, and this is especially true of sulfur-coated urea. The urea should be spherical and have a smooth surface. If the granules or prills have sharp corners, surface blips, cracks, dimples, or holes or if they are agglomerated so that the granules have a very irregular joint, they are much more difficult to coat completely and effectively with sulfur.

When the sulfur coating is sealed with wax, most imperfections in the granules can be fixed by using heavier coatings of sulfur and/or wax. However, this reduces the nitrogen content of the product and increases the manufacturing and shipping cost per unit of nitrogen. When the coating is only sulfur, the imperfections are much harder to correct. Wax can flow into dimples, cracks, holes, etc. to seal them, where sulfur cannot because it freezes on impact with the relatively cold substrate. For example, most prilled urea has a small dimple or hole in the surface of the sphere. Prills have been successfully treated with a conventional coating which includes wax, but efforts to make a sulfur-only product from prills have been discouraging. Products made from prilled urea with a 25% total coating (20% sulfur, 3% wax, 2% conditioner) had a 1-day dissolution rate of 22%. Using the same prills as a raw material, the sulfur-only product with a sulfur coating of 33% had a 7-day dissolution rate of 49%, and data indicated that increasing the sulfur coating would have little effect on the 7-day dissolution.

It appears that granules, regardless of their size, must have coatings of about the same thickness to produce similar dissolution rates. If this is a valid observation, then the amount of sulfur a substrate urea requires per unit of weight varies directly with the surface area of the urea substrate or inversely with the square of the average diameter of the granules, provided all other variables are constant. Therefore as long as agronomic benefits and handling characteristics are equal, if the size of the urea particles to be coated increases, the coating needed will decrease, reducing the production costs per unit of nitrogen and increasing the nitrogen content in the final products.

The urea substrate should be hard. Soft granules or prills give less support to the sulfur coating, making it more susceptible to fracture on handling.

The urea substrate may be precoated to prevent caking during shipment and storage without adverse effects on the final sulfur-coated product. In most pilot-plant sulfur-coating tests, the substrate material used was spray drum-granulated urea supplied with a 1% kaolinite coating of Barnet clay. Kaolinite or some other parting agent or a chemical caking inhibitor, such as formaldehyde, is needed in most cases to prevent the substrate urea from caking unless the sulfur coating is applied to very recently produced urea. A 1% coating of Barnet clay actually may lower the sulfur requirements by as much as two percentage points. However, the economics of including an extra step to precoat feed urea with kaolinite prior to sulfur coating is extremely questionable at best. Increasing the preconditioner kaolinite to more than about 1% is detrimental. Urea preconditioned with formaldehyde can be coated with sulfur without any apparent effect on the quality of the final product.

Fleming has developed a process where finely divided powder, in particular carbon black, is precoated on the urea to reduce substantially the sulfur used in the final coating. No work has been done in the TVA pilot plant on this aspect of sulfur coating (*11*).

Sulfur. TVA uses bright sulfur mined by the Frasch process containing about 500 ppm carbon and about 55 ppm ash. The carbon content of the sulfur appears to be a very important factor. Most of the carbon is in hydrocarbons which react with hot sulfur to form a solid carbon–sulfur complex, carsul. Because this reaction is very slow and continuous at normal storage and use temperatures, the carsul is difficult to keep removed from the sulfur. This will cause extreme problems if the sulfur is not handled properly or if there are excessive hydrocarbons in the sulfur.

Wax. A microcrystalline wax manufactured by a major U.S. oil company is used to seal the sulfur coating in the conventional process. Because of the energy crisis, many of the major oil companies have recently changed their product lines. Microcrystalline wax production has been curtailed by some companies and even eliminated by others. Thus, the availability and cost of microcrystalline wax is very uncertain, especially for new consumers. Therefore, TVA is continuing extensive small-scale work with sealants and has found other waxes and organic substitutes which appear almost as effective. At present, these products have been tested only on a bench scale, but pilot-plant tests are planned in the immediate future for the most promising ones.

Conditioner. The main function of the conditioner is to absorb the oils in the wax and thus remove the stickiness from the product. Therefore, it is needed only when an oily wax is used. Primarily, diatomaceous earth has been used as the conditioner because of its oil-absorbing capacity of 139 lb/100 lb as reported by the supplier. Kaolinite has a much lower oil-absorbing capacity of 34 lb/100 lb and was a poor substitute, requiring twice as much coating to provide an unsticky final product.

Basic Process

The basic process for producing a sulfur-coated urea with a wax sealant is shown in Figure 1. In the first step of TVA's continuous process, granular urea is heated in a rotating drum by an electric radiant heater. The heat prepares the surface of the granules to accept the subsequent sulfur coating. The urea granules are then sprayed with molten sulfur in another rotary drum. The sulfur can be sprayed through nozzles which use heated, compressed air as the atomizing force, or it can be atomized by pumping the sulfur at high pressures through hydraulic atomizing nozzles. In either case, a uniform coating of sulfur is built up on the individual particles. The material is discharged to a wax-coating drum by a bucket elevator. Wax (2% of total product weight) is applied to seal and fill the imperfections in the sulfur coating. The urea is then cooled to solidify the wax. A fluid bed cooler is used to minimize handling problems with the sticky, wax-coated material while cooling it efficiently. The material is discharged from the cooler to the conditioning drum where diatomaceous earth (2% of the total product weight) is applied. The product is screened to remove any oversize material formed

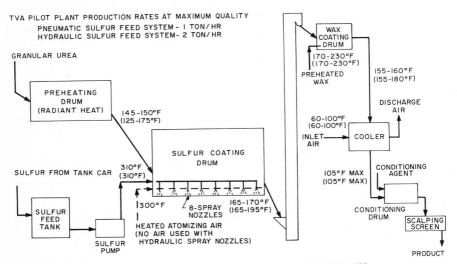

Figure 1. Flowsheet of process for coating urea with sulfur and a wax sealant using either pneumatic or hydraulic sulfur-spraying system

in the process. Usually about 0.1–0.5% of the total production is oversized in the form of agglomerates.

When producing a controlled-release urea coated only with sulfur, no subsequent processing is required, but some quick cooling, which is beneficial under certain operating conditions, is recommended.

Pilot-Plant Equipment

The TVA pilot plant was originally designed to produce 1 ton/hr of sulfur-coated urea, but it now produces 2 tons/hr. It consists of a urea storage hopper, urea feeder, urea preheater, sulfur feed system, sulfur-coating drum, wax-coating drum, fluidized bed cooler, conditioning drum, and auxiliary process equipment. The preheater is a rotary drum about 2 ft in diameter and 6½ ft long. Heat is provided by 12 440-V resistance radiant-type heaters installed inside the rotating drum. The temperature of the urea is controlled by cycling the heaters on a timed basis between one-quarter and full-load output. The shell of the preheater is smooth inside and insulated outside.

The sulfur-coating drum is about 4 ft in diameter and 6 ft long. The internals of the drum and the associated sulfur piping are completely different depending on whether the sulfur is atomized pneumatically or hydraulically. A schematic of sulfur-coating drum operation with the two types of sulfur spraying is shown in Figure 2.

When the pneumatic spraying system is used, the sulfur-coating drum has a 3⅜-in. retaining ring at the discharge end. It contains no flights and is insulated outside. It has a carryover shield to prevent occasional granules from falling on the steam-heated pipe and melting. The dripping melt causes agglomerates when it falls into the granule bed. The sulfur is distributed through a steam-heated header equipped with eight air-atomized spray nozzles. The nozzles are positioned to spray the sulfur directly down at the most active part of the rolling granule bed. The atomizing air for the sulfur sprays is steam heated in a shell-and-tube heat exchanger. The pneumatic sulfur-spraying system is fully discussed later in this chapter.

When the hydraulic spraying system is used, the sulfur-coating drum has a 5-in. retaining ring at the discharge end. Twenty-four lifting flights are installed in the drum at 15° intervals. The flights are straight with flat surfaces 3 in. wide and 58 in. long. They are installed parallel to the axis of the drum and, to increase their lifting capacity, are slanted 15° forward from the perpendicular with the drum shell. As the drum rotates, the flights carry material out of the granule bed and discharge it onto a collecting pan. The pan extends the length of the drum and is installed parallel to the axis of the drum and level with the horizontal with respect to its length. The pan is 20½ in. wide and is sloped counter to the direction of the drum rotation so that the granules will cascade down it. As the granules discharge from the pan, they strike a deflector plate, forming a continuous vertical curtain of free-falling granules on which the sulfur is sprayed. The slope on the collecting pan and its relative position with the bed of cascading granules can be varied. The sulfur is distributed through a steam-heated header equipped with eight hy-

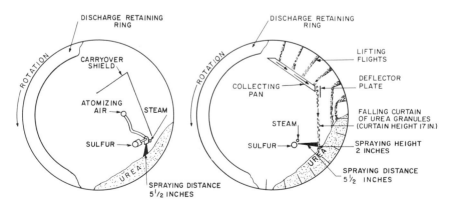

Figure 2. Sulfur-coating drum operation

draulic spray nozzles. The sulfur is atomized when it passes through the small orifice tips of these nozzles at high pressures.

The wax-coating drum is about 3 ft in diameter and 2 ft long. Wax is supplied to the wax-coating drum from a steam-heated tank with a small submerged diaphragm-type metering pump. It is distributed onto the bed of material in the wax-coating drum through a drip tube distributor. The material from the wax-coating drum feeds through a rotary airlock valve into the fluidized bed cooler by gravity.

The cooler has a bed area of 10 sq ft. Air for fluidization and cooling is provided by a centrifugal fan which forces the air through the cooler. The ductwork and damper arrangement allows close control of airflow into the cooler. The fluidized bed unit is very efficient, giving product temperatures that are within 5°F of the cooling air temperature at the 1 ton/hr rate. Pressure taps and temperatures gages are positioned to permit monitoring of air pressures and temperatures throughout the cooler and airflow system. Air from the cooler is vented through a stack. The cooled product flows by gravity through a rotary airlock valve to the conditioning drum.

The conditioning drum is about 4 ft in diameter and 3 ft long and has a smooth shell. A diatomaceous earth conditioner is metered to the conditioning drum with a vibrating screw-type feeder.

The conditioned product is discharged from the conditioning drum onto a belt conveyor which feeds a 4-mesh scalping screen. Product from the screen is fed by conveyor to a bagging hopper. Bucket elevators are used to fill the urea storage hopper and to elevate the product from the sulfur-coating drum to the wax-coating drum. All rotary equipment is driven with electric motors. All of these units have variable-speed drive except the conditioning drum. Instrumentation provides process data throughout the system. All of the pilot-plant processing equipment has performed well.

Specific Processes and Variables

The process and the related variables depend on whether the plant is producing sulfur-coated urea with or without a sealant and whether the sulfur is being applied with the pneumatic or hydraulic sulfur-spraying system.

Pneumatic Sulfur-Spraying System. SULFUR-COATED UREA WITH SEALANT. Figure 3 is a flow diagram of the pneumatic sulfur-spraying system. This system provides better atomization than the hydraulic system, which is discussed later. With the pneumatic system less sulfur is required, producing a higher analysis sulfur-coated urea fertilizer.

The sulfur is pumped out of a steam-heated, scale-mounted tank by a hydraulically operated diaphragm-type metering pump. Its head contains a stainless steel diaphragm and two ball check valves and is remotely mounted in the weigh tank. The sulfur in the tank is about 300°F. It is sucked into the pump through an 80-mesh wire cloth with an area of 60 sq in. to filter out any large particles of carsul or other impurities which might cause pumping or spraying problems. The steam-jacketed pipeline from the pump to the spray nozzles is used as the fine temperature control, adjusting the temperature of the sulfur to 310°F for spraying. Sulfur is at its lowest viscosity (0.004 lb/ft-sec) at 310°F.

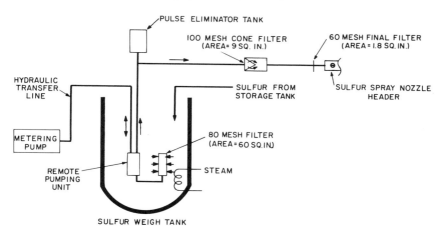

Figure 3. Flow diagram of system for supplying sulfur to pneumatic spray nozzles

The sulfur is pumped through two more filters before entering the sulfur header. A 100-mesh stainless steel cone filter with 9 sq. in. of filter area and a small 60-mesh filter with 1.8 sq in. of filter area are installed immediately ahead of the sulfur header. The larger unit contains most of the particles which might otherwise plug the orifices or nozzles. The small unit acts as a cleanup filter to retain anything formed in the pipeline between the large filter and the sulfur header or anything which might have passed the larger filter because of particle orientation or filter failure. Both of these filters need cleaning after about 3000 lb of sulfur has passed through them. The inlet filter is self-cleaning because of the pump pulsation. However, the carsul which accumulates in the bottom of the weigh tank must be drained occasionally. The sulfur must be adequately filtered if the process is to work. This may be the most important part of this process from an operational standpoint, and it is even more important in the hydraulic spraying system.

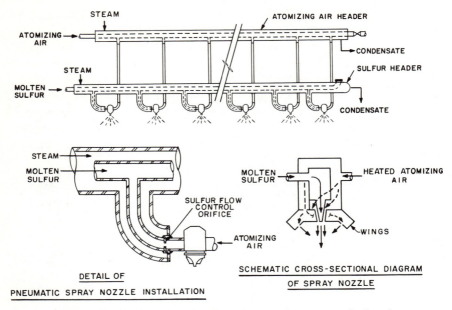

Figure 4. Sulfur spray header system and spray nozzle details

The filtered sulfur is metered equally to each spray nozzle by an in-line sulfur flow control orifice ahead of each atomizing nozzle. The development of this orifice control system to ensure uniform feed to the nozzles was a main feature of the successful design of the pneumatic coating system. The nozzles (*see* Figure 4) require little or no hydraulic pressure because the heated atomizing air, ≈ 295°F, acts as an asperating force and pressure drops in the nozzle are minimal. This, coupled with the physical characteristics of sulfur and slight differences between nozzles, caused starvation of some nozzles and excessive output on others when flows to the individual nozzles were not regulated. With identical orifices installed before each nozzle, they delivered the same sulfur flow, provided at least a 5–10 lb/sq in. pressure drop occurs across the orifice. Unless the orifices are extremely small, approximately 0.018–0.028 in. in diameter, this pressure drop cannot be obtained. This small opening and the somewhat larger opening of 0.052 in. in diameter at the nozzle tip necessitate the extremely fine filtration requirements mentioned above. Besides providing some asperating force, the air which passes through the annular space surrounding the liquid nozzle atomizes the sulfur as it expands on leaving the nozzle. The air to the wing tips of the nozzle only shapes the spray pattern of the atomized sulfur and has very little atomizing or asperating effect on the molten sulfur being sprayed.

Variables which affect the production rate and the quality of the coated products are discussed below.

1. Total coating weight of sulfur, wax, and conditioner. If all other variables are kept constant, the dissolution rate or slow-release characteristics of the final product are controlled entirely by the amount of sulfur and wax applied to the urea. Figure 5 shows the standard product quality that can be produced by the sulfur-coated urea pilot plant using a pneumatic sulfur-spraying system. The total coating requirement for a product with a desired dissolution rate can be estimated as well as the nitrogen and sulfur contents. For example, a product having a 30% dissolution rate in 7 days can be produced with about a 13% total coating, resulting in a product containing 40% nitrogen and 8% sulfur. The graph is based on the use of excellent quality granular feed urea coated with 2% wax and 2% conditioner. The urea from which this graph was constructed had been coated with 1% conditioner by the manufacturer to minimize caking.

2. Size and condition of the urea. This was previously mentioned but is re-emphasized because of its extreme importance. The substrate material graphed in Figure 5 was spherical, smooth, and closely sized. Its Tyler screen size is: +6 mesh = 0%, −6 +7 mesh = 9%, −7 +8 mesh = 70%, −8 +9 mesh = 19%, −9 mesh = 2%.

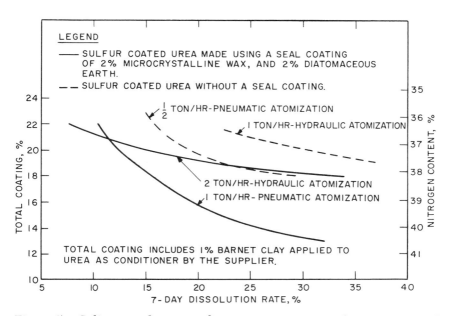

Figure 5. *Sulfur-coated urea quality using optimum production rates and operating conditions for pneumatic and hydraulic spraying systems*

3. Type and size of pneumatic sulfur spray nozzles. External atomizing nozzles are used as previously described and schematically pictured in Figure 4. The nozzles produce a finely atomized spray in the shape of an elongated oval. Depending on nozzles used, various alterations are needed to maximize the spraying efficiency and to eliminate operational problems associated with start-up and shutdown. The nozzles should be sized for maximum atomization at the spray distance of 4¾ in. while discharging a sulfur spray with a density of about 2.5 lb/sq in./hr at that distance.

4. Number of sulfur nozzles. Eight nozzles/ton of sulfur-coated urea/hr are considered optimum. More than eight nozzles/ton are not desirable from the standpoints of operation ease and design simplicity. As the ratio of nozzles-to-production rate is reduced, the product quality is decreased (*see* Figure 6). When the sulfur was coated with eight nozzles at the production rate of 1 ton/hr, a total coating of 13% (8% sulfur) produced a material with 7-day urea dissolution rate of 30%. When four nozzles were used at the same production rate, a total coating of 20% (15% sulfur) was required to produce a material with the same 7-day urea dissolution rate of 30%.

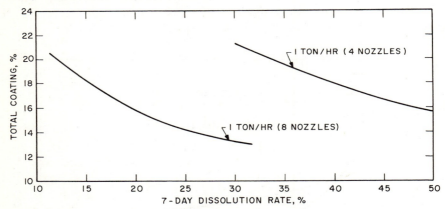

Figure 6. Effect of eight vs. four spray nozzles on the total coating requirement for sulfur-coated urea at the production rate of 1 ton/hr

5. Position of sulfur nozzles. For best results, the spray nozzle header is located parallel to the axis of the sulfur-coating drum. The nozzles in the header are 8 in. apart on centerline. The header is located so each nozzle is spraying vertically down 4¾ in. onto the fastest moving section of the rolling bed. If the nozzles are placed farther away, the sulfur dust loading in the drum is increased and the quality of the product is decreased. Increasing the spray distance beyond about 8 in. from the bed is both impractical and hazardous as sulfur explosions may occur if

dust concentrations exceed 35 g/cu m in air. Reducing the spray distance increases the sulfur spray density at the impact point and decreases the coverage area of the spray nozzle. Both are detrimental to the quality of the coating.

6. Rotational speed of the sulfur-coating drum. Increasing the interaction of particles in the sulfur-coating drum and the speed by which they pass under the sulfur spray nozzles increases the quality of the sulfur coating. With the 4-ft diameter sulfur-coating drum the best speed is 65–70% of critical speed (the speed at which centrifugal force holds the particles to the shell of the drum for its entire revolution). Beyond this speed, material carryover is high and sulfur–urea build-up on the coating drum accelerates. Normally, there is some build-up on the shell of the sulfur-coating drum from sulfur mist in the drum. In the pilot-plant unit this buildup provides a rough surface on the drum which facilitates the rolling action of the granules in the drum. The erosive action of the granules rolling and bouncing around keeps the build-up in check. However, in a full-scale plant a cleaning device may be needed for the drum shell.

7. Atomizing air volume. The atomizing air provides the force to atomize the sulfur. The maximum available air pressure of 80 lb/sq in. gage is used. The volume of air used at this pressure is approximately 4.7 standard cu ft/min (SCFM)/nozzle. This air is split so that the ratio of air through the wing tips of the nozzle to that which goes through the center opening is approximately 2.2:1.

8. Process temperatures. The process temperatures are critical. Without good temperature control the sulfur does not crystallize in the proper structure and the coating is less effective. However, because the product is sealed with wax, some variation in the sulfur structure can be tolerated, and the temperatures are somewhat less critical than in the sulfur-only process.

Temperature of preheated urea. The urea should be preheated to 145–150°F. If the urea is insufficiently preheated, the sulfur coating has a crusty appearance and high dissolution rates result from an otherwise well coated material. Much higher temperatures give trouble further along in the process.

Liquid sulfur temperature. As previously mentioned, the best temperature for sulfur spraying is approximately 310°F. Tests have shown that as the temperature decreases, the effectiveness of the coating is reduced. Temperatures up to 317°F are acceptable but above 318°F, the sulfur viscosity is so great that good sulfur atomization is virtually impossible.

Atomizing air. Air heated to the maximum temperature attainable in the pilot-plant system (290–300°F) is used to atomize the sulfur with good results. Somewhat higher temperatures of about 315°F are probably more desirable.

Sulfur-coating drum temperature. As the material passes through the sulfur-coating drum, its temperature increases because it receives the heat of fusion released by the solidifying sulfur. When a high oil microcrystalline wax is used, it is important that the wax be applied to granules at 155–160°F. The urea must, however, be above the freezing point of the wax for the roll-on distribution system to be effective. The wax must be able to flow onto the granules and into the imperfections before it freezes. Since the urea loses 5–10°F between the sulfur-coating drum and the wax-coating drum, the sulfur-coating drum exit temperature is controlled at about 165°F.

Liquid wax temperature. When using the wax referred to above, the preheat temperature is 220°F. The wax tempreature does not appear critical. When applied, the wax quickly adjusts to the temperature of the sulfur-coated urea granules because their mass is 50 times greater.

Cooling temperatures. The liquid wax applied to the sulfur-coated urea must be solidified before storage. Using the microcrystalline wax with a high oil content requires cooling the product to at least 105°F. Unless this is done, the wax remains too sticky and must be coated with excessive amounts of conditioner to prevent poor handling characteristics. As previously mentioned, a fluid bed cooler cools quite adequately without chilled air, provided the bed is fluid at all times and the material never comes in contact with the fluidizing screen. It is important that the cooling air temperature be no lower than about 60°F. At temperatures cooler than this, the wax freezes too fast and agglomerates are formed.

SULFUR-COATED UREA WITHOUT SEALANT. In order to make a slow-release urea fertilizer coated only with sulfur, the sulfur crystalline structure and uniformity must be closely controlled, for there is nothing to seal coating imperfections. However, several of the steps in the sealant process can be eliminated, and some of the other processing equipment can be substantially reduced in size. Since no wax or conditioner is added, all equipment associated with these steps can be omitted. Cooling is still important, but the cooling capacity can be substantially reduced. Handling complications of the waxed product are eliminated, greatly simplifying the engineering design of the cooler and associated product-handling equipment. On the negative side, more preheating capacity is needed, and the sulfur-coating drum throughput capacity is only half of that when a sealant is used because increased amounts of sulfur are needed. A process diagram of the pilot plant producing ½ ton/hr of urea coated only with sulfur is shown in Figure 7.

Variables 2 through 7 in the preceding section are more important than when a sealant is used, but their specified values remain the same. The total coating weight (Variable 1) is substantially changed. The dissolution rate of the product still varies with the amount of coating applied, but now the coating is only sulfur. About 50% more of it is required on an actual sulfur basis to give a specific dissolution rate.

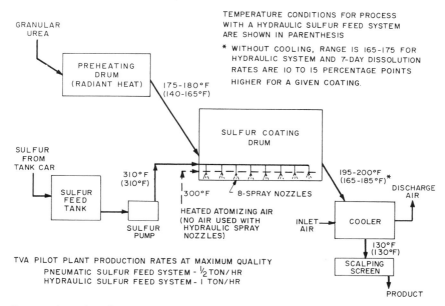

Figure 7. Flowsheet of process for sulfur-coating urea without a sealant using either the pneumatic or hydraulic sulfur-spraying system

However, the total coating of the product is only increased by 4½ percentage points (see Figure 5). A total coating of 18% (17% sulfur, 1% pre-applied clay conditioner) gives a 7-day dissolution rate of 30%. A total coating of only 13.5% is required to give a 7-day dissolution rate of 30% when wax sealant is used.

The sealant coating of 2% wax and 2% conditioner has a raw material cost of about $19/ton of nitrogen coated if it can be obtained. The additional coating of sulfur in the sulfur-only product has a raw material cost of about $9/ton of nitrogen coated. Thus, the sulfur-only coating is very competitive even though it does have a lower nitrogen content in its final product—38% as compared with 40% for sealant coating. The above figures were all based on using minus 6- plus 10-mesh substrate urea, producing a product having a 7-day dissolution rate of 30% and on January 1974 costs of sulfur, wax, and conditioner, f.o.b., Muscle Shoals, Ala.

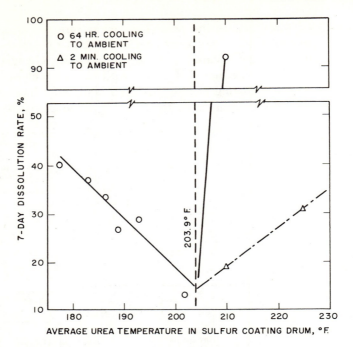

Figure 8. Effects of urea temperature in sulfur-coating drum on dissolution of sulfur-coated urea with no seal coating using the pneumatic sulfur-spraying process

The process temperatures (Variable 8) are very different from those used when applying a sealant except for the molten sulfur and its atomizing air which are still maintained at about 310 and 295°F, respectively. The most critical temperature is that of the substrate when the sulfur is applied. At no place in the sulfur-coating drum should the temperature of the substrate be allowed to exceed 204°F. Yet, the closer the temperature is to this value, the better the final product quality. Therefore, the preheater exit temperature is regulated to give a sulfur-coating drum exit temperature of 195–200°F, which is about as close as can be controlled in the pilot plant. At a production rate of ½ ton/hr in the pilot plant, the temperature of the material exiting the preheater or entering the sulfur-coating drum is approximately 175°F.

Figure 8 shows the effect of the substrate temperature on the quality of the final product. This figure also shows that if the temperature of the material is allowed to exceed 204°F in the sulfur-coating drum, then quick cooling, such as in the fluid bed cooler, will prevent a complete degradation of the protective coating. It has also been shown that some quick cooling is helpful on products where the temperature does not exceed 204°F. Therefore, a small cooler is recommended.

Hydraulic Sulfur-Spraying System. SULFUR-COATED UREA WITH SEALANT. Figure 9 is a flow diagram of the hydraulic spraying system used in the 1 ton/hr sulfur-coated urea pilot plant. Since nozzle openings are smaller (0.007–0.015 in.) than the orifices used in the pneumatic system, sulfur filtration in this system is even more important.

A 400-mesh screen is installed in the sulfur line from the weigh tank to remove carsul, ash, and other foreign materials. The sulfur then flows through a 2-μm filter to remove any small particles which might still be in the sulfur and which might combine to form larger particles further downstream. Sulfur is pumped by a double-acting air-driven piston pump capable of developing pressures up to 3000 psig. Because of the high temperatures and sulfur–hydrocarbon incompatibility, the pump is lubricated with a fluorocarbon lubricant.

All of the piping downstream from the sulfur pump is stainless steel high-pressure piping which was tested at 4500 psig. A pulse eliminator tank is located downstream from the sulfur pump. Air trapped in the top of the tank provides a constant pressure on the sulfur spray nozzles and thus maintains a constant flow during changes in the stroke direction of the pump. A 200-mesh filter is installed in the pulse eliminator primarily to take out trash which might get into the line during pump maintenance. Pump pressure is regulated by the amount of air pressure applied to the piston air motor. The sulfur flow rate is determined by the combination of this pressure and the size and number of spray nozzles installed in the sulfur header. The only way of reducing sulfur

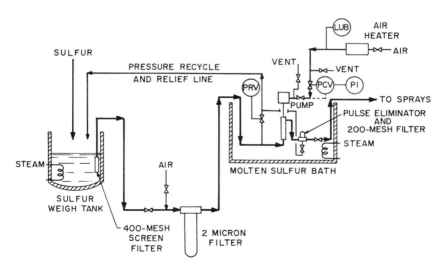

Figure 9. Flow diagram of system for supplying sulfur to hydraulic spray nozzles. All sulfur lines are steam jacketed or submerged in molten sulfur.

flow through a given spray nozzle or set of spray nozzles is to reduce the pressure in the hydraulic system.

To provide for depressurization of the system, a recycle line is connected between the pump and the sulfur weigh tank. A pressure relief valve is installed on the feed side of the pump to prevent possible pressurization of this portion of the piping should a malfunction of the pump allow backflow from the pulse eliminator tank. Air is provided for purging the sulfur line and for back-cleaning the 400-mesh screen in the sulfur weigh tank. The sulfur pump and all pipe fittings and valves not steam jacketed are submerged in a molten sulfur bath to keep sulfur from freezing in them during operating periods.

The spray header itself is much simpler than the spray header of the pneumatic system. No in-line flow control orifice is needed since each nozzle at these high pressures acts as its own flow control orifice. Furthermore, because no air is used, no pneumatic piping is needed. A high-pressure union allows the header to be rotated so that the angle of spray with reference to the bed can be easily changed. Also, the position of the nozzles with reference to the urea bed can be altered as desired by using externally mounted adjustable hangers. The entire header unit is steam jacketed.

Except for the absence of any pneumatic equipment, the simplified flow diagram for the hydraulic spraying system is the same as that shown in Figure 1 for the pneumatic spraying system. However, major changes as discussed previously were made to the inside of the sulfur-coating drum and to its operation to provide a curtain of falling urea granules. The sulfur is sprayed onto this curtain. The resulting product is not so high in nitrogen content as that made using pneumatic nozzles (see Figure 5), but the process is much more operational. A total coating of 18% gives a product with a 7-day dissolution rate of 30% containing about 38% nitrogen.

A primary reason for installing a hydraulic spraying system was to decrease sulfur mist and dust formation in the sulfur-coating drum. This has been accomplished. Based on an airflow through the sulfur-coating drum of 200 SCFM, dust loading in the atmosphere of the sulfur-coating drum ranged from 0.068 to 0.163 g/cu m/lb of sulfur sprayed when pneumatic nozzles were used. With hydraulic nozzles, the dust loading has been reduced 88–98%. The dust loadings have ranged from 0.003 to 0.008 g/cu m/lb of sulfur sprayed. Problems associated with the dust have been either substantially reduced or eliminated.

Safety. The minimum explosive concentration for sulfur is 35 g/cu m of air. The heaviest dust loading measured in the sulfur-coating drum while using hydraulic atomizing nozzles was 3.2 g/cu m of air, well below the explosive range.

Air pollution abatement. Dust loadings for best operating conditions were less than 0.006 g/cu m of airflow/lb of sulfur sprayed. Further cleanup of the discharged air, if required, appears quite easy, probably involving only a heated cyclone separator.

Sulfur-coating drum build-ups. Such build-ups appear to have been eliminated. A related problem of agglomerates caused by sulfur particles coalescing on the hot header piping with subsequent dripping of molten sulfur into the bed of urea has also been substantially reduced.

Variables which affect the hydraulic process are generally the same ones which affected the pneumatic process with some exceptions. However, the relative importance and optimum values of these variables are quite different in a number of cases.

1. Total coating weights of sulfur, wax, and conditioner. Provided all other variables are constant, the dissolution rate or slow-release characteristics of the final product are still controlled by how much sulfur and wax are applied. However, slight changes in coating content definitely affect the final product in the hydraulic process (*see* Figure 5). When sulfur was applied by the pneumatic spray process, a total coating reduction of 5 percentage points, 18.5–13.5%, increased the 7-day dissolution rate from 15 to 30%. With the hydraulic process, a reduction of only 2 percentage points, 20–18%, gave the same increase in 7-day dissolution rate, 15–30%.

2. Size and condition of the urea. There is increased mixing action in the sulfur-coating drum which is a result of the lifting flights and the falling curtain of urea. There is less particle segregation and thus more uniform coating between particles. Therefore, particle size and surface condition are not as critical with this system, although they are still extremely important.

3. Type and size of hydraulic sulfur spray nozzle. The hydraulic spray nozzles used are very simple. They have no vanes, cones, or other inserts which can cause problems. Sulfur is forced by pressure through an elongated oval making a highly atomized fan pattern. The nozzle tip size is varied depending on the amount of sulfur to be sprayed and the desired atomizing pressure. Changes can be made quickly and easily. A final protective filter is inserted into each nozzle assembly ahead of the nozzle tip (*see* Figure 10). The mesh size of the final filter varies depending on the size of the orifice in the nozzle tip. This filter normally removes very little material, and thus does not present a cleaning problem.

4. Hydraulic pressure. The pilot plant is able to spray at pressures up to 3000 psig. Theoretically, nozzle atomization efficiency increases with pressure. Therefore the most desirable conditions for sulfur application are the highest pressure attainable and the smallest nozzle available. However, filtration problems associated with the smallest nozzles com-

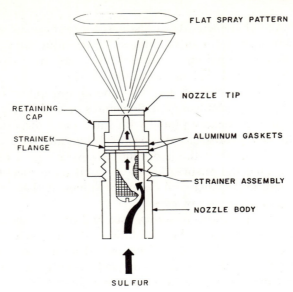

Figure 10. Diagram of hydraulic spray nozzle assembly

bined with the practical number of nozzles that can be used in the process make a nozzle with a tip opening of 0.011 or 0.013 in. the most desirable. These nozzles are operated at 750–1500 psig. Tests showed that pressure increases up to 3000 psig had very little effect on the final product using these nozzles, however the lower operating pressures are much more desirable from an operational standpoint.

5. Position of the sulfur nozzles. For best results, the nozzles are oriented to spray sulfur perpendicularly onto the falling curtain of urea at a point 2 in. above the bottom of the curtain. The distance between the nozzles and the curtain is 5½ in. The nozzles in the header are 7¾ in. apart.

6. Number of sulfur nozzles. The number of sulfur nozzles should be minimized. Four nozzles/ton of production/hr appear to give optimum results.

7. Rotational speed of the sulfur-coating drum. The rotational speed of the sulfur-coating drum still affects the coating quality. However, it has much less effect than with the pneumatic spraying system. The density of the falling curtain of urea and the mixing action in the drum are increased by higher drum speeds. However, the speed at which the granules pass in front of the nozzles (which is extremely important) is not affected by changes in drum speed in the hydraulic process.

8. Process temperatures. The process temperature control is less critical in the hydraulic process than in the pneumatic system.

Temperature of preheated urea. Data show that the urea needs to be preheated only to about 100°F for sulfur-coating purposes. However, material must be 160°F when coated with the microcrystalline wax. Therefore, for the best product, the urea should be preheated to at least 130°F.

Liquid sulfur temperature. The best temperature for the liquid sulfur is still 310°F.

Sulfur-coating drum temperature. The sulfur-coating drum temperature should not exceed about 200°F. Best results were obtained when the exit temperature of the sulfur-coated urea was maintained at 160–185°F.

Wax-coating temperature. Wax-coating temperatures do not appear to be so critical in the hydraulic process. Good results are obtained when the wax-coating temperature is 155–180°F.

Other temperatures and conditions not associated directly with the application of the sulfur and wax (e.g., cooling) remain the same as in the pneumatic spraying process.

SULFUR-COATED UREA WITHOUT SEALANT. When a slow-release fertilizer coated only with sulfur is made in the hydraulic spraying system, coating conditions are much different from those used with the pneumatic nozzle system. The basic flow diagram is the same except that, of course, that no atomizing air is used (see Figure 7). Operating variables 2, 3, 4, 5, and 7, discussed in the previous hydraulic spraying system section, are the same when producing a sulfur-only product. However, eight instead of four nozzles/production ton/hr (Item 6) will give better results. Figure 5 shows the product quality which can be produced at 1 ton/hr. Increasing the production rate to 2 tons/hr requires only increasing sulfur by 2 percentage points to maintain the same 7-day dissolution rate. As discussed previously, the same equipment when used in the pneumatic spraying system will only produce about ½ ton/hr of sulfur-only product or one-fourth the rate obtainable with the hydraulic process. The best product with a 30% 7-day dissolution rate made in the hydraulic system at 1 ton/hr is not so high in nitrogen content as the best product made with pneumatic nozzles at ½ ton/hr. However, at equal production rates of 1 ton/hr or higher, the hydraulically produced material is much superior.

Process temperatures in the hydraulic system are much lower than those in the pneumatic system. The urea should be preheated to at least 135°F for best results, but the temperature in the sulfur-coating drum should not exceed 175°F. At operating temperatures on either side of these values, the product quality falls off rapidly. Quick cooling of the final product to about 130°F is necessary to prevent some product degradation.

Sulfur-Coated Urea in the Future

Research work on the production and use of coated urea for controlled release of nitrogen is being continued. TVA hopes to reduce the ratio of nozzles to production rate and further decrease the size of the sulfur-coating equipment. A thorough agronomic evaluation of the material coated only with sulfur is being conducted which will fully indicate its potentials as a slow-release fertilizer. TVA plans to initiate demonstration-scale production of sulfur-coated urea in about 1975–76. Production rate for this unit will be about 10 tons/hr.

Literature Cited

1. Blouin, G. M., Rindt, D. W., U.S. Patent **3,295,950** (Jan. 3, 1967a).
2. *Ibid.*, **3,342,577** (Sept. 19, 1967b).
3. Blouin, G. M., Rindt, D. W., Moore, O. E., *J. Agr. Food Chem.* (1971) **19**, (5), 801.
4. Diamond, R. B., Myers, F. J., *Sulphur Inst. J.* (1972) **8**, (4), 9–11.
5. Engelstad, O. P., Getsinger, J. G., Stangel, P. J., "Tailoring of Fertilizers for Rice," *TVA Bull.* (1972) **Y-52**, 56 pp.
6. Allen, S. E., Mays, D. A., *J. Agr. Food Chem.* (1971) **19**, (5), 809.
7. Terman, G. L., Allen, S. E., "Leaching of Soluble and Slow-Release N and K Fertilizers from Lakeland Sand Under Grass and Fallow," *Soil Crop Sci. Soc. Fla. Proc.* (1970) 30 pp.
8. Shirley, A. R., Meline, R. S., *U.S. Pat. Off. Defensive Publication* **T912,014** (July 24, 1973).
9. "SCU: A Progress Report," *Sulphur Inst. J.* (1972) **8**, (4), 1–8.
10. McClellan, G. H., Scheib, R. M., *Sulphur Inst. J.* (1973) **9**, (3–4), 8–12.
11. Fleming, P. S., U.S. Patent **3,576,613** (Apr. 27, 1971).

RECEIVED May 1, 1974

4

Sulfur in Coatings and Structural Materials

T. A. SULLIVAN, W. C. McBEE, and D. D. BLUE

U. S. Department of the Interior, Bureau of Mines, Boulder City Metallurgy Research Laboratory, Boulder City, Nev. 89005

> *New uses for sulfur were developed to take advantage of a projected sulfur surplus. A spray material containing sulfur, talc, Fiberglas, and dicyclopentadiene was used to construct a block building. The blocks were surface-bonded together for structural stability by spraying with the mixture. Dirt-formed ponds were given an impervious lining by spraying with the material. The potential of low cost, chemically modified sulfur as a land stabilizer was evaluated. Dicyclopentadiene, dipentene, methylcyclopentadiene, styrene, and an olefinic liquid hydrocarbon were investigated as modifiers. A spray mixture of 7% dicyclopentadiene, 1% dipentene, and 92% sulfur effectively stabilized tailings from wind and rain erosion.*

The sulfur use program of the Federal Bureau of Mines was initiated in July 1972 when it became apparent that projected supplies of sulfur from secondary sources indicated an overwhelming surplus. During the past two decades supply and demand have been relatively closely balanced. The supply came largely from primary sources, although there was a large potential for sulfur from secondary sources during the past.

New environmental sources of sulfur will tap more of that potential. The amounts produced and the production starting times are speculative. The anticipated recovery of a substantial portion of the potential will seriously affect the sulfur industry. The Bureau of Mines forecasts that by the year 2000, 40 million long tons of sulfur could be recovered from coal, oil, gas, and metal sulfides. Domestic demands for the same year are forecast to be 30 million tons. Thus if only 75% of the potential from secondary sources were recovered, the demand for sulfur could be met without any production from primary sources.

An oversupply of most mineral commodities causes reduced prices, which, in turn, lead to increased consumption. Since normal sulfur uses are inelastic in this respect, a sulfur oversupply could not be relieved by increased conventional use. Therefore, the Bureau of Mines designed a program to develop new uses for the projected surplus sulfur. This paper describes the part of the program in which coatings and structural materials containing sulfur were developed.

Construction of a Building and Ponds with a Sulfur Spray Formulation

A sulfur-based coating material was developed by Southwest Research Institute to seal and support mine walls. The material was developed under a research contract with the U.S. Department of Interior, Bureau of Mines, for sulfur coatings for mine support (*1*). As part of the contract they also designed and furnished the Bureau with a portable spray unit for applying the material.

One of the first objectives of the sulfur use program was to determine if this sulfur spray formulation could be used in construction materials. Two methods were tested. First, a building was constructed by dry-stacking blocks and surface-bonding them by spraying with the formulation. About 10 yrs ago, Southwest Research Institute used a sulfur–Fiberglas formulation to construct a block building without mortar where the blocks were surface-bonded by brushing on the coating material (*2*). The new spray formulation would permit surface-bonding of the block by spraying. Second, the spray formulation was used in the construction of ponds or leach tanks to form an impervious lining. Such ponds could be used to contain acid leach solutions and to pond noxious wastes.

Experimental. Materials. The sulfur spray formulation contained 100 parts sulfur, 10 parts talc, 3 parts Fiberglas, and 2 parts dicyclopentadiene. The unit cost of the formulation was estimated at 2.3 cents/lb (*1*). The source and types of material are discussed later. To prepare the formulation, molten sulfur was allowed to react with the dicyclopentadiene as a plasticizer and was stirred at 130–150°C for at least 15 min. The Fiberglas was added for flexure strength, and the talc was added to give a thixotropic mixture and to disperse the glass fibers. When the mixture was homogeneous, it was ready for spraying. The material used to spray the inside walls of the building was held overnight at 135°C to minimize any objectional odor from unreacted dicyclopentadiene.

Equipment. The formulation was sprayed with the applicator built by Southwest Research Institute (*1*) for the Bureau. It consisted essentially of a propane-heated, oil-jacketed pressure tank with a mixer for blending the components of the spray formulation. Compressed air forced the spray formulation through an oil-heated spray hose. The building was surface-bonded with 300-lb batches, and the ponds were lined with 450-lb batches. The formulation was sprayed at temperatures ranging from 120 to 165°C. The optimum spraying temperature was in the range of 135–150°C. At lower temperatures, the formulation solidified without bonding to the subsurface, and at higher temperatures, the material was too viscous to spray. The tank was pressured at 15–30 psi.

Construction of Sulfur Building. A 10 × 20-ft building was designed to be constructed of commercial cinder block and consisted of two rooms divided by a cinder block wall. The building was erected on a concrete slab. The first course of block was laid and leveled on the slab. It was then bonded to the slab by pouring about 1/2 in. of the molten formulation into each block as shown in Figure 1. Then more cinder blocks were stacked to form the four walls (Figure 2). Because commercial cinder blocks are not perfectly square, it was necessary after the fourth course of block was laid to splash some of the molten formulation inside the blocks to keep the courses level and to stabilize the walls. At this point, the steel door frames were inserted into the walls.

Figure 1. Bonding the first course of block to the concrete slab

The spray formulation was used to grout the door frames to the cinder block as shown in Figure 3 and to grout the openings in one-half block on each side of the door openings. The window sills were formed with the formulation (Figure 4). It was also used to pour a sill over the door frames and to make lintel beams for the doors and windows by surface-bonding lintel blocks together with the formulation. For added strength, a 2-in. layer of the formulation was poured into each lintel beam over a 1/2-in. reinforcing bar. Figure 5 shows one of the lintel beams being inserted over a door opening.

Figure 2 (top). Dry-stacking the block to build the walls. Figure 3 (middle). Grouting the steel door frames to the block with the sulfur formulation. Figure 4 (bottom). Forming the window sill with the sulfur formulation.

Figure 5. Inserting lintel beam over a door opening

A room divider wall of cinder block was stacked after the outer walls were completed. The walls and the partition were then surface-bonded. Figure 6 shows a workman spraying the sulfur formulation to surface-bond the blocks. Both the inside and the outside walls of the building were sprayed, each with two coats of formulation.

After the building was surface-bonded, a 2 × 8-in. wood plate was bolted to the top of the walls. The bolts for the plate were set into the top of the wall by setting them in sulfur formulation poured in the top of some of the blocks. To accomplish this, newspaper was crumpled and inserted in the top block to a depth of 6 in. and the formulation was poured in around the bolt.

The ceiling of the building was constructed of steel decking. The decking was insulated with vermiculite which was held in place with a coating of the formulation. The spray machine was used to coat the

Figure 6. Spraying the sulfur formulation on the cinder block walls

Figure 7. Spraying the ceiling panels with the sulfur formulation

Figure 8. Completed sulfur-bonded cinder block building

vermiculite-filled ceiling panels with the sulfur formulation (Figure 7). The filled panels were then attached to the plate of the building to form the ceiling. A flat, shed-type roof was installed. Three sides of the building were painted with latex-based paint, and the fourth side was left exposed to observe the effects of weathering and aging on the formulation and the surface-bonding. The completed building is shown in Figure 8.

Results. The cinder block surface-bonded by spraying with the sulfur formulation gave a structurally sound building. The pertinent data for this process are given in Table I.

Table I. Specifications for Surface Bonding of Cinder Block Building

Sprayed surface	916	sq ft
Sulfur formulation sprayed	2400	lb
Average coverage	2.62	lb/sq ft
Average coverage, thickness	0.29	in.
Average spray time, 300-lb batch	15	min
Total spray time	120	min

Table II. Formulation Used in Cinder Block Building Construction (lbs)

Secure first course of block to slab	150
Prepare lintel beam	150
Seal door frames to block	200
Stabilize by splashing inside block	50
Grout block next door frames	450
Construct window sills and top of doorways	620
Set plate bolts	170
Spray vermiculite ceiling panels	300
Total	2090

In addition to spray surface-bonding the building, an additional 2090 lb of the sulfur formulation was used to construct the building. A summary of these uses is given in Table II.

Construction of Ponds. Three ponds were made by spraying the formulation on formed soil. Two small, shallow ponds of 65 and 350 gal capacity were constructed to determine the best method of applying the coating to obtain an impervious lining. A larger, double pond was then excavated and formed in the ground. The completed form for the double pond is shown in Figure 9. The double pond has a capacity of 3750 gal

Figure 9. Excavated form for the double pond

Figure 10. Spraying the double pond lining

and was designed so that the liquid from the upper pond overflows into the lower pond and is pumped back into the upper pond. The tops of the ponds, including the inside and outside edges, were sprayed first with 450-lb batches of the formulation mixed at about 145°C. Spraying time per batch was 15 min. After the tops of the ponds had been coated, the inside of each pond was coated (Figure 10). Three coats of the material were applied to give an average lining 0.46 in. thick. A total of 3600 lb of formulation was used. Before the final spray coating was applied, the lining was examined and any rough spots were sealed by hand-brushing with molten formulation. After the final coating was applied and had cooled to ambient temperature the ponds were filled with water and recycling was started as shown in Figure 11.

Results. The monolithic coating of the spray formulation applied to the double pond covered both the liquid-containing area, the top and outside edges, and also formed the spillway between the two ponds. After several days, some cracks developed in the coating at the top inside

Figure 11. Completed double pond

corners of the ponds. At first, these cracks were attributed to stresses caused by the allotropic transformation of the sulfur from the monoclinic to the orthorhombic form. Further observation led to the conclusion that the stress cracks were caused by expansion and contraction of the coating from temperature cycling through a range of approximately 50°C. While the formulation did bond to the soil, the shear strength of the soil permitted the lining to move. In the liquid area of the ponds, where temperature cycling was minimal, the lining did not crack. No cracking was noted in areas where the lining could move without restraint such as the outside of the ponds. Cracking occurred in major stress areas where the lining was not free to move such as in the upper inside corners of the ponds. Cracks which started in these areas finally extended into the liquid-containing area of the lining.

The ponds have successfully contained liquid for over 16 months. It was necessary to repair a few cracks in the liquid-containing area after 6 months by applying a strip of Fiberglas cloth over the crack and brushing on a layer of the spray formulation. No further evidence of cracking in these areas has been noted.

Discussion. The sulfur-bonded building has been in use for 1 yr. One room of the building was designed and used for a controlled temperature and humidity room to test various sulfur-coating formulations and to condition sulfur–aggregate and concrete test samples. The other room is used as an instrument control room for the sulfur demonstration area.

In addition to surface-bonding the block building, physical properties of the sulfur formulation were investigated. Previous work (1) had determined that the formulation is practically odorless on curing, that it is fluid enough to spray and will bond without dripping and that it is fire retardant.

To determine how long the sulfur formulation could be held in a molten state before spraying, its stability was measured on holding at 135 and 180°C. Stability was indicated by the hardness and impact strength of solidified samples taken at timed intervals. Molten formulation held at 135°C for over 2 months did not deteriorate. At 180°C, decomposition was noted after 30 days, and the material solidified into a black, friable mass after 45 days. This was attributed to carbonization of the mixture by dicyclopentadiene decomposition.

Impact strength *vs.* time was determined on solidified samples of the formulation that had been allowed to react from 30 min to 70 days at 135°C. The impact strengths were measured with a Gardner impact tester on 1/2 in. thick samples. Impact strength values for all samples stabilized in the range 7–9 in.-lb. Hardness of the solidified samples was about 75 Shore D after 1 hr and increased to a maximum of about 80

Shore D after 1 day. Material allowed to react for only 15 min before solidifying had an initial hardness of 60 Shore D and reached a maximum of 72 Shore D in 2 days.

Compression strengths of the solid formulation, determined by ASTM C39-49 test methods, were about 7000 psi. Flexure strengths of 1700 psi were determined by using ASTM C293-54T test procedures.

The spray formulation appears to be nontoxic because both fish and algae grow readily in the ponds. The pH level of the water in the ponds has remained constant, which indicates that bacteria are not converting sulfur to sulfuric acid. Samples of the formulation were exposed to bacterial attack in a controlled environmental chamber maintained at a high moisture level to accelerate the rate of attack. No evidence of bacterial attack on the samples has been detected after 1 yr of exposure under aerobic conditions. Initial corrosion tests showed that samples of the formulation did not lose weight after exposure in 1 and 5% sulfuric acid solutions for 4 months.

The success of the spray formulation as an impervious lining for leach or storage ponds requires provisions for its expansion and contraction from temperature cycling. Expansion joints could be used, or the formulation could be sprayed onto materials such as concrete, where it would bond with sufficient shear strength to prevent cracking from thermal cycling.

Sulfur Coatings for Dust and Water Erosion Control

One of the new uses suggested for sulfur was as a possible low-cost material for stabilizing copper tailings from wind and water erosion. These tailings comprise the bulk of mineral wastes that are not amenable to vegetative stabilization in the arid climate of the Southwest. The major problem is the control of dusting from the large areas of finely divided, dry mill tailings that become airborne with the slightest breeze. Chemical and vegetative stabilization of mineral wastes has been reported by Dean et al. (3, 4). Baker and Mallow (5) investigated the use of chemically modified sulfur as a dust pallative.

Initially ways of using sulfur alone as a stabilizing agent were studied. A molten sulfur coating was sprayed over the tailings. An application rate of about 25 tons of sulfur/acre initially stabilized the tailings. The coating did not penetrate the tailings but formed a layer on top which could be crushed by foot traffic. After exposure to the weather for about a week, the coating started to disintegrate. This probably happened because the sulfur both expanded and contracted from temperature cycling and stresses from volume contraction as the sprayed layer underwent allotropic transformation from the monoclinic to the orthorhombic form.

Other application methods were investigated, and the best method developed was to spray molten sulfur onto heated tailings. The surface of the tailings was heated to 120°C at a depth of ½ in. and sprayed with molten sulfur. The sulfur penetrated the top ½ in. and on cooling, formed a crust with the tailings strong enough to stabilize them against wind and water erosion. The crust was also strong enough to sustain light foot traffic.

The use of sulfur in tailings stabilization depends on its availability as a low cost material. Although the sprayed molten sulfur stabilized the tailings, the preliminary heating was both energy consuming and expensive. Chemical sulfur modification to give properties essential for direct spraying stabilization was also studied. The modifiers investigated were restricted to low-cost, commercially available unsaturated hydrocarbons to keep the costs within reason. The hydrocarbons chosen were those that would react with molten sulfur to form plasticized sulfur with either increased flexure strength or penetrability into the tailings. For this discussion, the term plasticized sulfur refers to a mixture of polymeric polysulfides and liquid S_8 as described by Currell and Williams (6).

Laboratory Penetration Tests. Bench-scale laboratory penetration tests were performed to evaluate the possibility of applying molten sulfur directly to soil test specimens consisting of fine construction sand to stabilize them from wind and rain erosion. The emphasis was on penetration of the sulfur or chemically modified sulfur into the test specimen directly, thus avoiding the necessity of applying external heat to promote penetration.

Molten sulfur, both pure and chemically modified, was poured onto the sand contained in 400-ml glass beakers. The average minimum penetration depth of the solidified sulfur into the sand was measured directly.

Dicyclopentadiene, dipentene, styrene, CTLA polymer, and methylcyclopentadiene dimer were used as modifiers. These modifiers are all available in commercial quantities at 5–12 cents/lb. All contain unsaturated double bonds suitable for direct reaction with sulfur. The materials used in this investigation were from the following sources:

> sulfur, 99.9% pure flake, Montana Sulfur and Chemical Co.
> talc, Mistron vapor talc, Cyprus Mining Corp.
> Fiberglas, No. 630 1/8-in. milled fiber, Owens Corning Fiberglas Co.
> dicyclopentadiene, dicyclopentadiene petrol oil, Velsicol Chemical Corp.
> dipentene, Glidco dipentene 300, Gliddens–Durkee
> olefinic liquid hydrocarbon, CTLA polymer, Exxon Chemical Co.
> methylcyclopentadiene dimer, industrial grade, Exxon Chemical Co.
> styrene, Baker Grade, J. T. Baker

Sufficient reaction time was allowed in each test to insure complete reaction between the particular modifier and sulfur. With modifier concentrations in the range of 1–19 wt %, the reaction times ranged from less than 10 min at 220°C to over 120 min at 120°C. Combinations of two modifiers, namely, dipentene–dicyclopentadiene and dipentene–styrene, were also tested.

Results. Pure elemental sulfur applied between 120 and 160°C hardly penetrated into construction sand. Dicyclopentadiene–sulfur mixtures produced negligible penetration at concentrations of 3–9% at 120–160°C. Above 9%, the penetration depths ranged from 3 to 11 mm at 130–160°C. Above 140°C, the dicyclopentadiene–sulfur reaction is exothermic and extremely difficult to control. When uncontrolled, the material was extremely viscous and the penetration was negligble.

Methylcyclopentadiene dimer results were identical to those for dicyclopentadiene at 120°C. Because of these results and the close structural similarity of the modifiers, further tests were discontinued with this modifier.

Styrene–sulfur mixtures produced penetration depths of 3–5 mm at 3–15% concentration and 120–130°C. There was negligible penetration at 140°C and above because of the drastic viscosity increase of the mixture.

CTLA polymer–sulfur mixtures produced negligible penetration at 120–140°C regardless of concentration. All mixtures with CTLA polymer were extremely viscous. Above 140°C, the mixtures decomposed and evolved considerable hydrogen sulfide gas.

Dipentene–sulfur mixtures produced negligible penetration at 1–15% concentration and 120–200°C. Above 15% the penetration depths were 7–17 mm at 140–180°C.

Dicyclopentadiene–dipentene–sulfur and styrene–dipentene–sulfur mixtures produced comparable results. Penetration depths were 1–3 mm at 140–160°C for styrene or dicyclopentadiene concentrations of 4–7%. From 7 to 10% styrene or dicyclopentadiene, the penetration depths were 3–6 mm at 140–160°C. The dipentene concentration was varied from 1 to 3% and essentially reduced the viscosity. The penetration depth did not vary with changing dipentene concentration.

Laboratory Reaction Tests. It was necessary to understand the reaction parameters for the modifiers dipentene, styrene, and dicyclopentadiene to aid future field stabilization tests. Laboratory tests were performed to define the typical character of the material in the liquid and solid state. The conditions necessary to form completely plastic noncrystalline sulfur with each modifier were also established.

The tests were performed using approximately 200-g batches. Modifier concentration was varied from 1 to 33 wt % and reaction times were from less than 2 min at 200°C to greater than 70 hrs at 120°C. Each specimen was allowed enough time to react completely at the given composition and temperature. The specimen was then poured into a small rectangular aluminum mold and allowed to solidify.

The viscosity observations were qualitative and were made by directly comparing resistance to flow upon pouring the particular liquid into the molds. The viscosity of each mixture was compared with the viscosity of elemental sulfur at 120°C, which was considered low.

The flexibility observations were also qualitative and were obtained by physically bending the cast bars until they fractured.

Crystallinity, or degree of plasticization, was checked by visual and microscopic examination of both the cast and fractured surfaces. These observations were confirmed by observing the etching character produced by the action of carbon disulfide on the specimen. If the material was crystalline, the crystalline faces and boundaries were uniformly delineated. If the sample was plastic, no delineation occurred.

Results. The results obtained from the laboratory reaction tests are summarized in Table III.

Table III. Laboratory Reaction Test Results

Temp (°C)	Concentration (w/o)	Character of Material	
		Liquid State, Viscosity	Solid State[a]
Dicyclopentadiene			
120–140	>13	low	NC brittle
140–180	>13	high	NC brittle
120–140	<13	low	MC brittle
140–180	<13	high	MC brittle
140–180	> 6	high	NC brittle
140–180	< 6	high	MC brittle
Dipentene			
120–200	>26	low	NC flexible
120–200	<26	low	MC brittle
Styrene			
120–140	>13	low	NC flexible
140–150	>13	high	NC flexible
120–140	<13	low	MC flexible
120–140	<13	high	MC flexible

[a] NC = noncrystalline; MC = mixed crystalline, noncrystalline.

Field Tests. A series of nine test plots were sprayed with modified sulfur mixtures which were based on the laboratory penetration and reaction test results. Rectangular test plots 6 × 7¼ ft (1/1000 acre) were built and filled with fine construction sand. Before spraying with the molten sulfur, half of each plot was textured by raking to produce a

Figure 12. All-electric sulfur spray machine

hill and valley effect. A new 400-lb capacity spray machine (Figure 12) was designed for these tests. Better temperature control and simplified spraying were achieved by electrical heating and mixing instead of using propane gas for heating and a gasoline engine for mixing. It is capable of operating at up to 200°C and 40 psi. Data for the test plots are given in Table IV, and the finished plots are shown in Figure 13. Small rectangular test specimens were cast from each spray batch before applying the material to the test plots. These specimens were evaluated for degree of plasticization, flexural strength, and mode of fracture. On some batches, the short-term aging effects on the flexural strength were determined.

Table IV. Spray Plot Data

Plot	Spray Composition (w/o)[a]	Reaction Time (min)	Reaction Temp (°C)	Spraying Temp (°C)	Wt Used (lb)
1	94 S, 1 DP, 5 DCPD	75-DP 15-DCPD	150-DP 160-DCPD	160	49
2	93 S, 1 DP, 6 DCPD	150-DP 45-DCPD	160	160	53
3	92 S, 1 DP, 7 DCPD	125-DP 65-DCPD	160	160	39
4	85 S, 15 styrene	30	120	120	110
5	78 S, 14 styrene, 8 DP	60-styrene 75-DP	120–190-styrene, 190-DP	190	48
6	87 S, 13 styrene	80	160	160	66
7	88 S, 12 DCPD	90	120–160	160	53
8	93 S, 7 DCPD	150	160	160	31
9	96 S, 4 DP	30	180	180	50

[a] DCPD = dicyclopentadiene, DP = dipentene, S = sulfur.

Figure 13. Stabilization test plots

Additional test specimens were cut from the smooth and textured areas of each plot and evaluated for penetration depth, flexural strength, bonding, and mode of fracture.

Results. Flexural strength values for specimens taken from each spray batch, before spraying, are given in Table V. Flexural strength values obtained after short-term aging of specimens from five of the spray batches are plotted in Figure 14. Values obtained with sulfur specimens are also included. Typical fractures illustrating the extremes of failure observed with different materials are shown in Figure 15. Data from the evaluation of test specimens cut from the nine spray plots are given in Table VI. Figure 16 illustrates typical penetration cross sections from the spray plots.

Discussion. Chemically modified sulfur was superior to elemental sulfur for direct penetration into soil test specimens. Except for the CTLA polymer, the modifiers promoted penetration.

In general, sulfur penetration into soil test specimens depended on both temperature and modifier concentration. Using styrene or dicyclo-

Table V. Spray Formulation Data

Spray	Composition	Flexural Strength (psi)	Fracture Mode
1	94 S, 1 DP, 5 DCPD	360	cleavage
2	93 S, 1 DP, 6 DCPD	425	cleavage
3	92 S, 1 DP, 7 DCPD	1540	cleavage
4	85 S, 15 styrene	175	intergranular
5	78 S, 14 styrene, 8 DP	25	intergranular
6	87 S, 13 styrene	525	cleavage
7	88 S, 12 DCPD	1160	cleavage
8	93 S, 7 DCPD	955	cleavage
9	96 S, 4 DP	145	intergranular

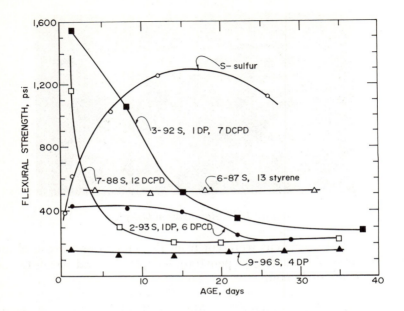

Figure 14. Short-term aging effects on flexural strengths

pentadiene as a modifier, these two variables were self-defeating. As indicated in Table III, the higher the temperature ($> 140°C$) and the concentration, the greater the viscosity of the mixture. With dipentene mixtures, viscosity did not increase. For this reason the dipentene mixtures appeared to be the most promising. One per cent dipentene can be used as a viscosity reducer, and the styrene or dicyclopentadiene could be used in concentrations suitable to achieve penetration at the desired temperature.

Table VI. Spray

Plot	Spray Composition	Penetration Depth (mm)	
		Smooth	Textured
1	94 S, 1 DP, 5 DCPD	3	3
2	93 S, 1 DP, 6 DCPD	3	4
3	92 S, 1 DP, 7 DCPD	4	6
4	85 S, 15 styrene	1	1
5	78 S, 14 styrene, 8 DP	12	13
6	87 S, 13 styrene	3	4
7	88 S, 12 DCPD	3	4
8	93 S, 7 DCPD	2	4
9	96 S, 4 DP	4	5

[a] All specimens aged more than 30 days before testing.

Figure 15. Fractured sulfur specimens—cleavage (top) and intergranular (bottom) (2×)

The increased penetration for modified sulfur over elemental sulfur can be attributed to several factors.

1. When modifiers are used, the freezing point of sulfur is lowered and in some cases supercooling occurs, allowing greater penetration time.

2. Modifiers retard or prevent crystallization. With pure sulfur, nucleation of crystals is instantaneous upon contact with the soil test specimen.

3. Modifiers allow higher temperature application without increased viscosity, allowing greater penetration time.

Plot Data

Strength (psi)[a]		Material Character– Color and Crystallinity[b]	Fracture Mode
Smooth	Textured		
505	470	yellow NC	cleavage
680	820	tan NC	cleavage
1140	1340	tan NC	cleavage
250	405	yellow C	intergranular
<50	<50	brown C	intergranular
715	990	yellow NC	cleavage
865	860	tan NC	cleavage
215	335	brown NC	cleavage
330	535	yellow C	intergranular

[b] NC = noncrystalline; C = crystalline.

Dipentene mixtures allow suitable penetration by using a minimum modifier concentration (< 10%). These mixtures can be held at elevated temperatures without increasing the viscosity.

Sulfur can be fully plasticized by using the modifiers dipentene, styrene, and dicyclopentadiene. Sulfur can be plasticized with dicyclopentadiene at two minimum concentrations, as indicated in Table III: 13% at reaction temperatures of < 140°C and 6% at reaction temperatures > 140°C. This effect probably results from cracking of the dicyclopentadiene dimer molecule, which doubles the molecules available for reaction. The higher percentage dicyclopentadiene mixture was initially flexible. Upon aging, both plasticized materials became brittle. The reaction is exothermic and very difficult to control above 140°C. When uncontrolled, extreme viscosity increases were encountered.

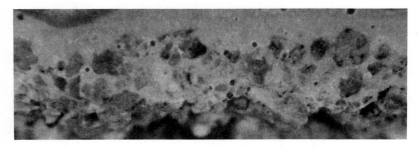

Figure 16. Penetration cross section of a typical spray plot (6×)

Styrene reacts readily with sulfur, and plasticization requires a minimum concentration of 13%. The plasticized material has remained flexible for 3 months after solidification. At 140°C and above, styrene is a very effective viscosity increaser for sulfur.

Dipentene reacts slowly with sulfur, and a minimum concentration of 26% is required for complete plasticization. The plastic material has remained flexible 3 months after solidification. Dipentene is a very effective viscosity reducer above 160°C.

Flexural strength values obtained from the various spray mixtures can be used to evaluate the extent of reaction. As indicated by the values shown in Table V, the flexural strengths of spray batches 4, 5, and 9 were extremely low. Here the reaction time was not long enough, and, therefore, plasticization was not achieved. Spray 9 was intentionally applied before it was fully reacted. Here the modifier concentration was held to a minimum and was used only for viscosity control. Complete reaction was obtained in the other mixes.

Examination of the fractured surfaces yields a direct correlation with flexural strength. The higher strength mixes and elemental sulfur all fractured by a cleavage mode whereas the low strength mixes failed by intergranular separation.

Short-term aging results indicated that the flexural strength of sulfur mixes modified with dicyclopentadiene significantly decreases over the testing period. Signs of flexibility are generally absent after 1 day. The flexural strengths of styrene- and dipentene-modified sulfur remained essentially constant over the short-term aging period.

Sulfur, upon aging, tends to increase in flexural strength, as shown in Figure 14. Because of the extremely brittle nature of sulfur, reliable flexural values are difficult to obtain after 15 days. This undesirable property of sulfur points out the necessity of chemical modification to prevent or retard these brittle properties.

The penetration depths measured on the spray plots were comparable with the laboratory tests. The mode of fracture on specimens cut from the plots was the same as that observed on the specimens taken before spraying. Sulfur-to-sand bonding was sound on all spray plots.

Flexural strengths generally increased after spraying because of the aggregate effect produced when the molten material penetrated into and bonded with the sand. Texturing generally improved penetration. As a result, the textured regions generally exhibit greater flexural strength.

Weathering has produced considerable cracking in plots 4 and 8. Slight cracking has occurred in plots 1, 5, 6, 7, and 9. Plots 2 and 3 remain completely sound.

Chemical modification of sulfur with 6–7% dicyclopentadiene and 1% dipentene at 160°C was the most suitable mixture tested for spraying a coating on tailings to prevent wind and rain erosion.

Summary

A sulfur spray formulation containing sulfur, talc, Fiberglas, and dicyclopentadiene was successfully used to construct a block building. The dry-stacked cinder blocks were surface-bonded for structural stability by spraying with a thin coating of the molten formulation. The same formulation also formed an impervious lining on a dirt-formed pond.

The chemical modification of sulfur to alter its physical properties for use in land stabilization was investigated. Commercially available, low-cost unsaturated hydrocarbons were used as chemical modifiers. By using modifiers alone or in combination, the properties of sulfur in both the molten and solid state were controlled, permitting its use by either direct penetration or thin crust formation to stabilize tailings. The most suitable formulation was sulfur modified with 6–7% dicyclopentadiene and 1% dipentene.

Literature Cited

1. Dale, J. M., Ludwig, A. C., "Sulfur Coatings for Mine Support," Final Report U. S. Dept of the Interior, Bureau of Mines, Southwest Research Institute, **11-3124,** Contract H0211062, Nov. 1972.
2. Dale, J. M., "Sulphur-Fibre Coatings," *Sulphur Inst. J.* (1965) **1,** (1), 11–13.
3. Dean, K. C., Havens, R., Harper, K. T., "Chemical and Vegetative Stabilization of a Nevada Copper Porphyry Mill Tailing," Bureau of Mines **RI 7261,** 14 pp, May 1969.
4. Dean, K. C., Havens, R., Glantz, M. W., "Methods and Costs for Stabilizing Fine-Sized Mineral Wastes," Bureau of Mines **RI 7896,** 26 pp, 1974.
5. Baker, E. J., Jr., Mallow, W. A., "The Use of Sodium Silicate and Sulphur as a Dust Palliative," Contract Report **S-69-1,** Contract No. DA-22-079-end-482, U.S. Army Material Command, Corps of Engineers, Southwest Research Institute, May 1967.
6. Currell, B. R., Williams, A. J., "Plasticization of Sulfur," *Int. J. Sulfur Chem.* (1972) **2,** 245.

RECEIVED May 1, 1974. Reference to specific companies, brands, and trade marks is made for identification only and does not imply endorsement by the Bureau of Mines.

5

Sulfur in Construction Materials

M. A. SCHWARTZ and T. O. LLEWELLYN

Tuscaloosa Metallurgy Research Laboratory, U.S. Department of the Interior, Bureau of Mines, Tuscaloosa, Ala. 35401

> *Two research developments have been made by the U.S. Bureau of Mines on the use of sulfur as a mortar to produce prefabricated brick panels and the use of foamed sulfur–gypsum mixtures to produce lightweight wallboard. In the first development, the excellent adhesive and strength characteristics of sulfur were applied to prefabrication construction technology. In the second, a novel process was developed that combined simultaneous foaming and surface compressing to produce a paper-covered wallboard, both less dense and stronger than conventional plaster board. Sulfur exhibits numerous advantages as a construction material such as low cost, high availability, excellent physical properties, good chemical compatibility, and good processability, but potential problems of flammability and reaction product odors must also be considered.*

The relatively low cost, high availability, and excellent physical properties of sulfur have justified considering it as a construction material. Its excellent bonding or adhesive qualities as well as chemical compatibility with many types of inorganic materials make it competitive for many applications which currently use cement or plaster binders. Its water repellancy makes it even more attractive for construction applications.

The lower costs for sulfur, resulting from its recent production from sour gas, have made it even less expensive than commonly used portland cements. This is particularly so in the northern parts of the country where Canadian imports have a significant transportation advantage. Industry in this country is also developing its own stack gas sulfuric acid production brought about largely by air pollution controls.

While the advantages of using sulfur in construction materials are indeed impressive, the disadvantages must also be considered. Energy, in the form of heat, is required to melt the sulfur (mp ~ 115°C). Sulfur burns readily in air (ignition point at atmospheric pressure: 248–261°C) and additives to make it fire retardant or coatings possibly to make it fire proof are required. Sulfur also readily reacts with many organic materials, creating unpleasant odors. It should be possible to circumvent flammability or odor problems, and energy required to melt sulfur is far less than that required to produce most cements.

The studies reported here were conducted by the Tuscaloosa Metallurgy Research Laboratory as part of the Bureau of Mines research program to develop new uses for sulfur, especially in the high volume construction industry. Freeport Sulfur Co. dark type elemental sulfur containing 0.22% carbon and bright type containing 0.02% carbon were both used in this investigation. The study is not an exhaustive investigation and was made primarily to determine the merit of using sulfur in sulfur-bonded prefabricated brick panels, sulfur–sand mixtures, and lightweight sulfur–gypsum wallboard.

Prefabricated Brick Panels

The concept for this development was based on problems currently existing in the prefabricated brick panel industry which in recent years has had an impressive upsurge of interest in both Europe and the United States. These brick panels are constructed 10 ft high by 12 ft wide or even larger at factories and transported to construction sites. While these panels are being used successfully, basic problems which are inherent to the mortar include relatively long curing times and marginal strengths which would require reinforcements.

Using sulfur as the mortar could circumvent these problems and enhance the use of prefabricated brick panels. Molten sulfur cures almost instantly on cooling and achieves high strengths in minutes. The compressive strength for sulfur is greater than 3000 psi as compared with only 1000 psi for conventional mortar. In addition, bad weather, often a deterrent to normal brick-laying practice, would not adversely affect the use of sulfur mortar. Sulfur curing would not be influenced by freezing temperatures as occurs with cementatious mortar, and sulfur would not be adversely influenced by rainfall because it is water repellant.

Sulfur mortar could also be used to produce entire brick walls on the ground to be installed in the manner of tilt-up concrete walls. Only the forms in which the bricks are set and in which the molten sulfur is cast would differ. To demonstrate the concept, the wood form shown in Figure 1 was constructed large enough to hold six sub-size clay bricks. These bricks were actually 3½ in. long by 1¾ in. wide by 1 in. thick. The bricks were evenly spaced by inserting 3/16-in. diameter wire at

Figure 1. Wooden form containing brick panel bonded with sulfur mortar

the bottom of the form and between each brick. Other spacer materials could also be used. Bricks were placed face down so that the wire spacers would produce a recessed joint. All inserts and sides of the form that would be exposed to the molten sulfur were coated with a petroleum grease to prevent adherence.

In these experiments the dark sulfur was heated to 140–160°C; the viscosity of the sulfur is lowest in this range. The molten sulfur was transferred by ladle from the melting pot and poured into the cavities between the bricks to a height approximately ¼ in. below the top of the bricks. The sulfur flowed easily, following the countours of the bricks closely to form a continuous network among the six bricks, and bonding them tightly together.

After cooling about 2 min, the form was removed, and the panel shown in Figure 2 was easily removed. Any rough mortar joints were

Figure 2. Sulfur-bonded prefabricated brick panel

smoothed down, using sandpaper. No objectionable odors were emitted at any time during processing or after construction. Although the flammability of this type of construction was not tested, fireproofing paints could be used on the back side of the panels.

The interface between sulfur and brick is shown in Figure 3. The bond is largely mechanical, with the sulfur closely adhering to the rough, porous brick surface. The chemistry of the system however has not been investigated, and the bonding may be a combination of chemical and mechanical forces. The bricks used in this study were somewhat unusual, consisting of 70% recycled waste glass and 30% natural clay (1). These

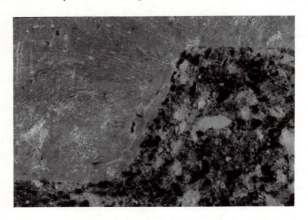

Figure 3. Sulfur–brick interface showing mechanical bonding (9×)

special bricks, even though less porous than conventional brick, offered no adherence problems.

Sulfur–Sand Mixtures

While considerable work was conducted by Southwest Research Institute (2, 3) to develop sulfur–aggregate mixtures, tests were also conducted on the effects on compressive strength of adding sand to sulfur to determine its utility as a mortar material. Two mortar sands of different particle size distributions were used with the sulfur to produce 1-in. cube compression test specimens. Particle size distributions are presented in Table I.

Test specimens were prepared by melting bright sulfur containing various amounts of sand to approximately 145°C. The slurry was then hand-mixed and poured into greased metal molds. To eliminate concavity resulting from nonuniform shrinkage, extra sulfur mixture was poured into each mold. On cooling, the excess material was easily

Table I. Particle Size Distribution of Mortar Sands

Mesh Size	Screen Fraction (wt %)	
	Coarse Sand	Fine Sand
Minus 20—Plus 35	3	—
Minus 35—Plus 48	21	—
Minus 48—Plus 65	35	15
Minus 65—Plus 100	30	31
Minus 100	11	54
Total	100	100

trimmed off with a knife, and rough surfaces were sanded smooth. A polished surface of a 50% sand specimen is presented in Figure 4. Test specimens were aged for 1, 10, and 30 days prior to measuring their strength. For comparison, cursory compression tests were conducted using 1-in. cube specimens in a hydraulically operated testing machine. The test specimen in Figure 5 indicates a typical brittle material type failure.

Compression test results are presented in Table II. Each strength value is the average of three tests which were generally in close agreement. Compositions with less than 50% sulfur were weaker because of insufficient matrix relative to the amount of sand. This was even more so with fine sand because of its greater surface-to-volume ratio. The higher sand-content compositions were less uniform as the sand tended to settle while the sulfur was molten. The finer sand did not settle as much as the coarser material. Aging did not significantly effect strength values.

For comparison, compression tests also were conducted on dark sulfur with no sand additions. Strengths of 2900 psi were obtained with

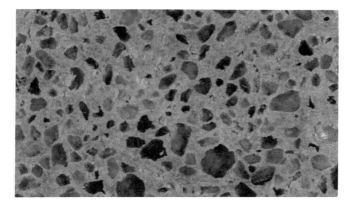

Figure 4. Polished surface of sulfur mortar containing 50 wt % sand (11×)

Figure 5. Sulfur–sand compression test specimen, after test

the dark sulfur as compared with 3800 psi exhibited by the bright sulfur. No explanation was apparent.

Figure 6 presents a comparison of the sulfur–sand mixture tests. Sand additions increase the compressive strength of the sulfur up to approximately 50 wt %. Based on these curves it would appear that 40–50 wt % of sand would produce the highest strengths.

Lightweight Sulfur–Gypsum Wallboard

Molten sulfur was developed into a lightweight wallboard material because it can be foamed. Just above its melting point of 115°C sulfur viscosity decreases to a minimum at approximately 160°C. Above 160°C the viscosity increases rapidly (4). To foam the sulfur at the lower part of this temperature range, additives are required which prevent the foamed structure from collapsing. Additives such as phosphorus pentasulfide and styrene monomer have been developed by Southwest Research

Table II. Compressive Strengths of Bright Sulfur–Sand Mixtures

Sand Addition (wt %)	Compressive Strength (psi)			
	Coarse Sand			Fine Sand
	1 Day	10 Days	30 Days	1 Day
0	3800	—	—	3800
20	—	—	—	4700
30	—	—	—	5600
40	6600	6600	6500	5900
50	6400	6600	7500	6900
60	5700	6500	5900	2900
70	4600	5600	5600	2100

Institute for producing rigid sulfur foams (5, 6). Foaming is produced by gaseous reaction products from the addition of calcium carbonate and phosphoric acid. Talc is added to stabilize the foam structure.

The foaming process was simplified by a new technique involving a waste gypsum product in place of the calcium carbonate–phosphoric acid additives. This waste material from the phosphate industry contains approximately 21% water and transforms to the hemihydrate form at 130°–150°C. At these temperatures, approximately 75% of the water volatilizes, creating the foaming action. A typical formulation used to produce foam, using either dark or bright sulfur, is (in parts by weight): sulfur, 200; phosphorus pentasulfide, 6; styrene, 6; talc, 15; gypsum, 5.5.

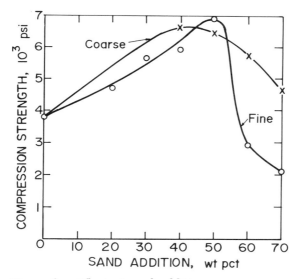

Figure 6. Effect of sand additions on compression strength of sulfur aged 1 day

Using this formulation, a process was then developed to produce lightweight wallboard. The surfaces of the foamed board were compressed or densified to produce relatively high smoothness. Paper sheets normally used to cover wallboard were applied to both sides simultaneously with the compression action. The excellent adhesive character of the sulfur produced tight bonding, and no special treatment of the paper was required.

A typical procedure producing 9-in. square test panels is:

1. The sulfur is first melted in a pot at 150–155°C.
2. Viscosity increasers, such as phosphorus pentasulfide and styrene, are added and permitted to react.

3. Inert materials, such as talc or clay, are added and thoroughly mixed.

4. Gypsum is added last with rapid mixing as foaming occurs rapidly.

5. The stable foam is cast or poured into a mold at a volume slightly greater than that of the mold. Suitable facing paper has been positioned on the upper and lower parts of the die which are transferred to the wallboard on pressing.

6. Pressure is applied to the foam, compressing it to the desired thickness.

7. The foam is cooled to below 120°C, at which temperature it solidifies and is handled without difficulty.

A simplified wood mold is used, and the mold surfaces are greased to prevent sticking. Construction materials may be made of metal, wood, or polymers.

The foamed structure, shown in Figure 7, is relatively uniform. The paper facing and compressed foam layer beneath the paper is evident.

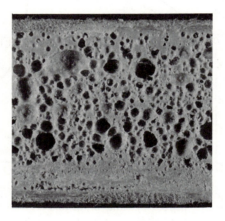

Figure 7. Cross section of sulfur wallboard structure (7×)

A typical sulfur wallboard sample had a density of approximately 0.5 g/cc as compared with 0.9 g/cc for plaster board. It was also evident that the sulfur board was considerably stronger than the plaster board even though strength tests were not conducted. Plaster board was easily broken by hand whereas much more physical exertion was required to break the sulfur board.

A section of sulfur wallboard is presented in Figure 8. It is able to accept nails and screws, which is important to this application. While tests have not been conducted on the noise and thermal insulating characteristics of the sulfur board, there is no reason to doubt their adequacy.

Figure 8. Foamed sulfur wallboard showing acceptance of nails and screws

One could visualize the possibility of self-bonding wood paneling to the foamed sulfur board the same way as paper.

Figure 9 is a schematic of a proposed continuous process in which paper facings and foamed sulfur are carried on a conveyor through compression rollers, a cooling device, and a cut-off machine, ready to be packaged and shipped.

As indicated earlier, the unpleasant odors created by organic reaction products pose a serious problem, especially with sulfur wallboard because it would be used in a confined area inside a building. The foam may be made rigid by other means which do not create unfavorable reactions. Another approach is to use hollow glass beads or expanded lightweight aggregate bonded by the sulfur instead of foaming the sulfur. This would eliminate the need for the phosphorus pentasulfide–styrene additives which are one of the causes of the odor.

Discussion

Although this investigation was preliminary, the Bureau of Mines has, nevertheless, demonstrated two new potential uses for sulfur—a mortar to produce prefabricated brick panels and a foamed lightweight

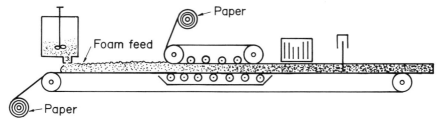

Figure 9. Schematic of proposed continuous sulfur wallboard production process

wallboard. The compressive strength of the mortar produced using equal amounts of sand and sulfur was 7500 psi after 30 days. The foamed sulfur wallboard exhibited good structural strength as compared with conventional wallboard while weighing much less.

Literature Cited

1. Tyrrell, M. E., Feld, I. L., Barclay, J. A., "Fabrication and Cost Evaluation of Experimental Building Brick From Waste Glass," *Bur. Mines Rep. Invest.* (1972) **7605**.
2. Dale, J. M., "Determination of the Mechanical Properties of Elemental Sulphur," *Mater. Res. Stand.* (1961) **1**, 23–26.
3. Ludwig, A. C., "Utilization of Sulphur and Sulphur Ores as Construction Materials in Guatemala," U.N. Rep. No. **TAO/GVA/4** (July 14, 1969).
4. "The Sulfur Data Book," Freeport Sulphur Co., McGraw–Hill, New York, 1954.
5. Dale, J. M., Ludwig, A., "Solid Foamed Sulphur and Process for the Manufacture Thereof," U.S. Patent **3,337,355**, 1967.
6. Dale, J. M., Ludwig, A., "Investigation of Lightweight Sulphur Foam for Use in Field Applications," TR 227, Contract **DAAG-23-68-c-0001**, U.S. Army Terrestrial Sciences Center, Dec. 1968.

RECEIVED May 1, 1974. Reference to brand or trade names is made for identification only, and endorsement by the Bureau of Mines is not implied.

6

Sulfur in Asphalt Paving Mixes

R. A. BURGESS and I. DEME

Shell Canada Limited, Toronto and Oakville, Ontario, Canada

> *The addition of molten sulfur to hot-mix asphalt paving materials increases mix workability so that the mixes may be placed without roller densification. As the mix cools, the sulfur solidifies and imparts such a high degree of mechanical stability to the mix that high quality mixes may be produced from poorly graded aggregates and even one-sized sands. The fatigue life and impermeability of these mixes is also increased. Sands are particularly suitable because of their availability and low cost in many areas. Sand–asphalt–sulfur mixes, comparable in quality to asphalt concrete, may be used as road bases and surfaces. Certain unique characteristics make them especially suited for applications such as pavement leveling courses, patching materials, impervious surfacings, castings, hydraulic applications, and others.*

The development of asphalt paving mixes using sulfur is the objective of an extensive program by Shell Canada Limited. This would permit the use of one-sized sands and poorly graded aggregates which, when mixed with asphalt and sulfur, form a material equivalent or superior in quality to asphalt concrete mixes made with dense graded aggregates (1, 2, 3, 4).

The prospect of using sulfur in hot mixes was examined in the 1960s by Shell to develop stable yet workable mixes. At mixing temperatures the molten sulfur contributed to the fluids content of the mix, but at temperatuers below its solidification point the sulfur added to the strength of the mix. In 1964, full-scale paving trials were carried out in Oakville, Ontario and St. Boniface, Manitoba in which conventionally graded aggregates and poorly graded sands were used with asphalt and sulfur. These early field trials provided the following knowledge which influenced the direction of our development program.

1. Solid or liquid sulfur could be readily mixed with aggregates and asphalt in a batch-type asphalt hot-mix plant.

2. Compacting the mixes by conventional rolling damaged the sulfur crystal structure, which formed as the mix cooled, causing poor subsequent performance of the rolled mixes.

3. Mixes which were not rolled performed very well.

4. Readily available, inexpensive one-sized sands could replace relatively expensive dense-graded aggregates in sulfur–asphalt mixes.

5. Problems with thickness control could be expected when relatively sloppy mixes were placed with a conventional asphalt spreader.

A sharp increase in price and decrease in availability of sulfur reduced the experimental effort temporarily, but laboratory work based on the above findings continued. Pourable paving mixes were developed containing one-sized sands which could be cast in place without rolling, much as portland cement concrete is handled. Based on satisfactory laboratory findings, a test road was constructed in Richmond, British Columbia in 1970, where a sand–asphalt–sulfur mix was cast between forms (5). The success of this trial, coupled with a decrease in the price of sulfur and the forecast for a long-term world sulfur surplus, led us to initiate an extensive research and development program to exploit sand–asphalt–sulfur mixes as road base and surface materials.

Our current program consists of two main parts. First, the laboratory program defines in detail the handling and engineering properties of aggregate–asphalt–sulfur mixes over a range of materials and formulations. Second, the laboratory findings are substantiated by evaluating the performance of the materials in pavement structures under service conditions. The performance results obtained from these basic field trials, along with supporting laboratory data, will establish the durability characteristics of sulfur–asphalt mixes and allow the definition of suitable pavement thickness design practices.

The versatile characteristics of sand–sulfur–asphalt mixes suggest a number of diverse possible uses for these materials (3). Some of these have already been investigated while others will be examined in the future. Various features of the sand–sulfur–asphalt process are covered by patents and patent applications in a number of countries. This process will be made available to industry under license when development work has been successfully completed.

The Function of Sulfur in Sand–Asphalt–Sulfur Mixes

Adding enough molten sulfur to hot aggregate–asphalt mix increases its fluidity and produces a mix which can be molded or shaped. Casting these mixes requires little or no consolidation effort, and, specifically with pavements, rolling is not required. It has been demonstrated by micro-

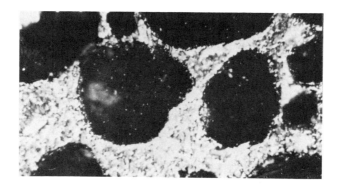

Figure 1. Photomicrograph of a polished sand–asphalt–sulfur mix surface

scopic studies that when the mix cools, the sulfur solidfies in the void spaces between the asphalt-coated aggregate particles, conforming to the configuration of the void. Sand–asphalt–sulfur mixes were prepared with translucent Ottawa sand, and from them polished sections were taken for optical micrographs. The sand particles in the typical optical photomicrograph of a polished mix section in Figure 1 appear black because of the asphalt film coating, which is visible from the underside of the particles. The same specimen area was examined with a Cambridge scanning electron microscope Kevex x-ray energy dispersive analyzer (Figure 2). The white area is sulfur which has intimately filled the interstices between the coated sand particles.

Graded aggregates are essential in conventional asphalt concrete mixes so that the interstitial spaces between the larger particles are filled

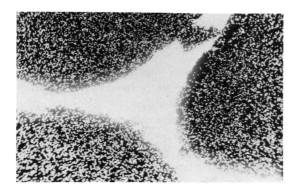

Figure 2. X-ray analyzer plot of a sand–asphalt–sulfur mix with sulfur indicated by white band

with smaller particles. However, in mixes prepared with sulfur, the sulfur performs as a conforming filler particle formed *in situ* and interlocks the aggregate particles giving the mix high mechanical stability. This diminishes the aggregate quality requirements, enabling high quality mixes to be designed using inexpensive materials such as one-sized sands. The asphalt film continues to exercise its conventional role as a visco-elastic binder, imparting flexibility and fatigue resistance to the mix.

Laboratory Program

In this program the properties of mixes in the hot, fluid state and the cold, hardened condition were examined. Although satisfactory mixes have been produced in the laboratory with a range of aggregates, the program was carried out with mixes produced using a medium-coarse sand. The results of the laboratory program are summarized in the following sections. More details on mix properties are given in Ref. 4.

Properties of Hot Aggregate–Asphalt–Sulfur Mixes

Workability. Besides giving strength to the cooled mixes, sulfur also provides workability to the hot mix so that it can be placed at the desired density without roller compaction. In conventional asphalt paving mix formulations, the asphalt content of the mix is limited by stability considerations. The asphalt imparts cohesiveness to conventional asphalt mixes, but the optimum asphalt volume is insufficient to provide adequate hot mix workability which permits mix densification without considerable rolling effort. By adding sulfur, however, the fluids content of a mix is increased, reducing or eliminating the compactive energy requirements for consolidating the mix, without adversely affecting its ultimate stability. The resulting mix workability may be used as a design criterion, and mix densification requirements may be more easily matched with proper placement equipment of suitable consolidation effort.

The laboratory results presented in Figure 3 illustrate that compactive energy only slightly increases the density of a high workability sand–asphalt–sulfur mix. Such a mix will flow and tend to assume its most dense configuration naturally. In contrast, considerable energy must be expended to densify the stiffer mix. For the mixes in Figure 3, the sand is fairly one-sized, and the mix is densified by forcing out air.

The results from full-scale experimental road projects in Canada have confirmed the above findings. When sand–asphalt–sulfur mixes were placed with vibratory screed finishers, increased vibratory energy had no significant effect on the densification of fluid mixes, whereas noticeable increases in density with corresponding increases in vibratory energy were observed with stiffer mix formulations.

Similar considerations should be remembered when placing graded aggregate–asphalt–sulfur mixes. In this case, however, some vibration is desirable to consolidate all mixes, regardless of consistency, so that the particles are oriented in the most dense configuration.

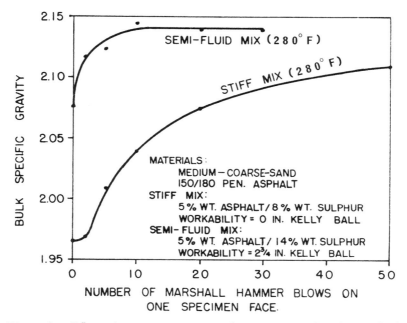

Figure 3. *Effect of compaction on mix densities for stiff and semi-fluid mixes*

From 300°F (150°C), the maximum mix processing temperature for averting fume emission (2, 4, 5), to 230°F (110°C), the crystallization point of sulfur, the sulfur is a liquid with a viscosity of 8–10 cP. For a 150/200 pen asphalt the viscosity ranges from about 160 cP at 300°F to 350 cP at 230°F. The variation in sulfur mix workability throughout this temperature range is largely related to the variation in the asphalt viscosity. The change in workability with temperature of aggregate–asphalt–sulfur mixes must be considered in mix processing, handling, transporting, and placing.

The workability of aggregate–asphalt–sulfur mixes is an important mix design criterion, and a measure of workability is necessary to assess the ease of placing a mix and to determine the consolidation energy requirements for densifying the mix. The cement concrete consistency tests for measuring slump (ASTM Test Method C 143 (6)) and ball penetration (ASTM Test Method C 360 (6)) are useful for measuring the consistency of aggregate–asphalt–sulfur mixes. However, all apparatus which comes in contact with the mix must be heated to prevent mix cooling and the resulting sulfur solidification on cool surfaces.

In the cement concrete industry the recommended practice is to consolidate low slump mixes by vibration and high slump mixes with hand tools. For example, ASTM Method C 192 (6) specifies consolidating test specimens by vibration if the slump is less than 1 in. Relationships between workability and ease of placing sand–asphalt–sulfur mixes are being evaluated on various field projects to assess the workability requirements for mix placement with various types of equipment.

Aggregate and Sulfur Segregation. Proper mix handling is essential to avoid separating the coarse aggregate from the finer mix material. The low consistency aggregate–asphalt–sulfur mixes are particularly liable to segregate, and precautions must be exercised during mix transport and discharge to ensure mix homogeneity.

Sand–asphalt–sulfur mixes are not subject to particle segregation because of uniform size distribution. These mixes may be transported and handled readily. Sands are, therefore, the best materials for use in paving mixes, considering their low cost in many areas, their ample availability, and their potential for developing high strength in sand–asphalt–sulfur mixes.

Molten sufur may tend to migrate within certain mixes, analogous to bleeding in cement concrete, so it is essential to exercise care in mix design to minimize seepage. A test method is currently being developed to measure sulfur seepage rates in mixes.

Properties of Hardened Mixes

Mix and Specimen Preparation. Most of the mixes described in this paper were prepared from a fairly one-sized medium-coarse sand with 35% voids in the mineral aggregate and with only 2½% of the material finer than No. 200 mesh. The sand grading is included in Table I. A 150/200 penetration grade asphalt was used.

The mix materials and the specimens were prepared as specified by ASTM Test Method D 1559 (7), except that initial wet mixing with asphalt was followed by a second wet mixing cycle with molten commercial grade sulfur. Marshall specimens were prepared with only two hammer blows on one specimen face. This light compactive effort was selected to expel the pockets of entrapped air and to prevent honeycombing along the side of the mold. Specimens for other tests were sawn from slabs which were produced by pouring the mix into rectangular molds and rodding around the edges to prevent honeycombing.

Marshall Stability. A factorial study was made to determine the Marshall stability for mix compositions with asphalt contents of 2–10% and sulfur contents of 0–20 wt % of mix. The mix stability of the two-

Table I. Aggregate Gradation

Cumulative Per Cent Passing

ASTM Sieve Size	Fine Sand	Medium-Coarse Sand	Coarse Sand	Low Stone Gravel	High Stone Gravel
3/4 in.				100	100
1/2 in.				95	90
3/8 in.			100	90	85
No. 4			98	85	70
No. 8		100	89	75	65
No. 16		97	74	59	55
No. 30	100	80	46	36	34
No. 50	75	48	20	16	11
No. 100	20	14	6	9	6
No. 200	5	2.5	2	7	5

blow specimens was measured according to ASTM Test Method D 1559, and the results are presented in Figure 4. The major findings of the study are as follows.

1. The sand–asphalt mixes had virtually no measurable stability.
2. The sand–asphalt–sulfur mix stability increased to a peak value with increasing sulfur content for all asphalt levels.
3. The mix stability decreased with increasing asphalt content for all sulfur levels.
4. Extremely high mix stabilities were attained by using sulfur compared with mix stabilities for dense graded asphalt concrete mixes which seldom exceed 2000 lb (8900N).
5. Sand–asphalt–sulfur mix stabilities were adequate even with excessively high asphalt contents, e.g., 10% asphalt.

On the basis of these observations, we concluded that using sulfur in an asphalt mix permits design for a predetermined mix stability value. This approach is unique in the realm of asphalt mix design because the stability level of conventional asphalt mixes is governed within narrow bounds by aggregate grading. Because the use of sulfur in mixes provides more freedom in the use of materials, it allows the designer to select the most appropriate stability level to optimize mix performance for a specific use.

Flexibility and Fatigue Resistance. The data presented in Figure 4 indicate that a given mix stability level may be attained using a variety of bitumen contents with the corresponding sulfur contents. For example, 1000-lb stability mixes may be made with asphalt contents of 2–10%. The major component influencing mix flexibility is the asphalt content. Thus, a lean mix composition, with only 2% asphalt, would be very brittle whereas the mix with 10% asphalt would be quite flexible and able to conform to roadway deformations without cracking.

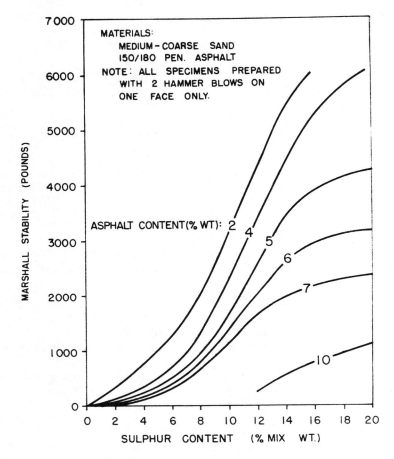

Figure 4. Effect of sulfur content on mix stability

The resistance of a pavement to fatigue under continuous flexing by traffic loads is important to the long-term performance of a pavement. The influence of sulfur content on mix fatigue life is illustrated in Figure 5. The mixes were prepared using the sand described in Table I with 6% bitumen and various amounts of sulfur. The tests were performed in a constant stress mode using a three-point bending apparatus similar to that described in Ref. 8. The results for tensile flexural strains were 1.5×10^{-4} and 2.0×10^{-4}, indicating that fatigue life increases to a peak with increasing sulfur content and then decreases.

In conventional asphalt concrete mixes, for which stability is a foremost consideration, the design asphalt content is generally close to the optimum asphalt content for fatigue life. In aggregate–asphalt–sulfur mixes the significant mix properties should be considered independently in establishing the optimum mix formulation for its intended use. As neither Marshall properties nor any other mix property reflect the fatigue properties of sand–asphalt–sulfur mixes, the designer is faced with either assessing the mix fatigue properties on the basis of previous experience or evaluating them directly.

Impermeability. Most naturally occurring sands are sorted materials which tend to be one-sized and have high voids contents. Generally, sufficient mix workability, stability, and other mechanical properties may be attained without completely filling the voids in the sand. This results in mixes with air voids often in excess of 10%. Test results, both in the laboratory and in the field, have shown however that such mixes are not necessarily permeable to water. It is believed that the sulfur network throughout the mix tends to isolate the air voids, thereby governing mix impermeability.

Considerable testing was done in the laboratory to evaluate the permeability properties of sand–asphalt–sulfur mixes. A constant head air permeameter, similar to the apparatus described in Ref. 9, was used. The coefficient of air permeability criterion below which mixes are con-

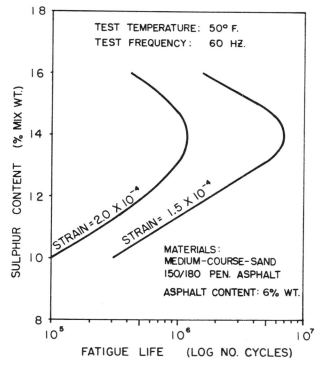

Figure 5. Relation between sulfur content and fatigue life for a sand–asphalt–sulfur mix

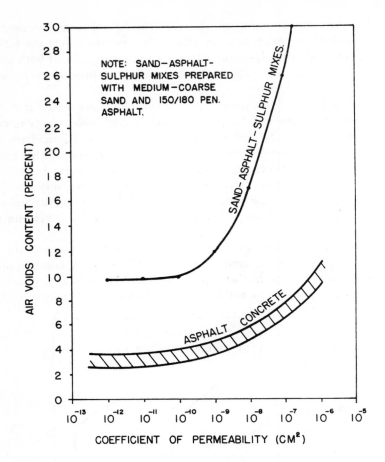

Figure 6. Relation between air voids content and permeability for mixes

sidered impervious is approximately 10^{-8} cm^2. On this basis, the test results in Figure 6 indicate that the maximum permissible air void content for sand–asphalt–sulfur mixes is 16%. The corresponding value quoted in literature for dense-graded asphalt concrete mixes is only 6%. Therefore, no specific attempt is made to fill the voids in mineral aggregates in sand–asphalt–sulfur mixes to reduce air voids contents to the same degree as in dense-graded asphalt mixes.

Aggregate Suitability Evaluation. As indicated earlier, most of the mixes studied were made with medium-coarse sand. However, a number of mixes were prepared to explore the range of aggregate gradations which may benefit from sulfur addition to asphalt compositions. The gradations in Table I cover materials from fine blow sands to coarse pit run gravels. The mixes were prepared in the laboratory with 5% of 150/200 penetration grade asphalt and 4–16% sulfur. No attempt was made to establish the optimum asphalt content requirements for each aggregate. Marshall specimens were prepared using the two-blow standard.

The Marshall stability test results in Figure 7 indicate that all of the mixes with low sulfur contents were unacceptably unstable. The stability of all of the mixes increases with increasing sulfur content. The medium- and coarse-sand mixes benefited the most from sulfur addition. They exhibited higher Marshall stabilities at lower sulfur contents than other mixes and at higher sulfur contents exceeded 2000 lb. (8900N), the usual upper limit attainable for high quality asphalt concrete mixes.

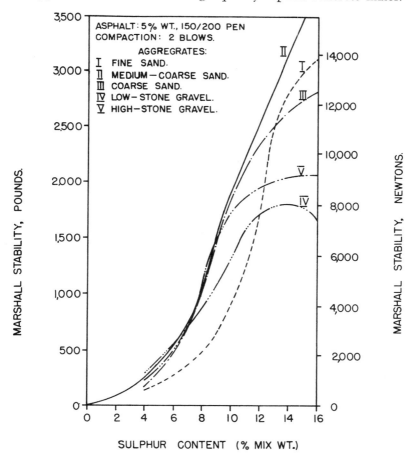

Figure 7. *Effect of sulfur content on Marshall stability for normal process mixes prepared using various aggregate gradations*

The study indicated that high stability mixes may be attained with aggregates ranging in gradation from fine sands to coarse gravels. However, the abundance of inexpensive sands makes them the most attractive candidates for mix production in most areas.

Field Program

Paving Trials. A number of sand–asphalt–sulfur test pavements were constructed in Canada without rolling the mixes. The pavements were designed so that mix durability and field performance variables could be evaluated under actual in-service conditions.

In 1970, 1300 ft of sand–asphalt–sulfur base was poured 4 and 6 in. thick between forms in Richmond, British Columbia (5), to assess the performance of the material as a building platform for a road over a very weak subgrade. In addition, 300 ft of a badly rutted street was overlaid using a conventional Barber-Greene finisher. A single mix composition was used in both cases.

In 1971, a three-quarter mile surface overlay was placed in Oakville, Ontario using thirteen mix formulations. Skid resistance, wear, and permeability have been measured periodically to assess the suitability of sand–asphalt–sulfur mixes as surfacing materials.

One thousand feet of test pavement was constructed in 1972 in St. Antoine, Quebec to evaluate the durability of the material under the prevailing low temperature and traffic conditions. A single mix formulation was used and placed 1½ and 3 in. thick.

A 1-mile test road was constructed in Tillsonburg, Ontario in 1972 to evaluate the performance of several mix compositions as a pavement base. Four mix compositions were placed as base 3½, 5½, and 7½ in. thick to evaluate the performance characteristics of the material under heavy traffic conditions.

The performance of the above pavements is being evaluated, and although the results are promising, a longer test period is necessary before conclusive results are obtained. A further field trial will be conducted at McLean, Saskatchwan during 1974. This trial will compare the performance of various sulfur–asphalt compositions and structures with conventional asphalt structures under low temperature conditions. The test road will include appropriate instrumentation.

Mix Handling Trials. The general objective in conducting field trials was to use, with minimum modification, conventional asphalt equipment for processing, transporting, and placing sand–asphalt–sulfur mixes. To date, the mixes have been processed in batch-type hot-mix asphalt plants. Details of the processing steps were published in a previous report (4). In 1975, we intend to evaluate a continuous-type hot-mix asphalt plant for producing sand–asphalt–sulfur mixes. Other types of plants may also be considered.

Experience gained in the field trials already conducted has provided guidelines for necessary transporter and spreader modifications. During 1973, in cooperation with some equipment manufacturers, we conducted a number of small field trials to evaluate modified equipment. The results have been promising, and the properly modified equipment should be available for future trials.

Potential Uses for Sand–Sulfur–Asphalt Mixes

Pavement Structure Bases and Surfacing. Sand–asphalt–sulfur mixes may be used in the construction of all types of pavements or for overlaying existing road structures. As the mixes are cast in place without roller compaction, they are suitable for road widening or bridging weak spots in the subgrade.

Sand–asphalt–sulfur surface-wearing courses prepared with coarse sands have a sharp, sandpaper-like surface texture. Skid resistance tests carried out up to speeds of 50 mph have given favorable results. The road surface is not susceptible to polishing because as soon as a sand grain is worn away or dislodged, another sharp sand grain is exposed. Fine sands, such as dune sands, are not suitable for riding surfaces because they yield surfaces which are too smooth.

Sand–asphalt–sulfur mixes may also be designed to be highly impervious. These mixes may be used for cement concrete bridge deck surfacing to reduce salt attack and corrosion of the bridge deck reinforcing.

The sand–asphalt–sulfur mixes are particularly suited as road bases and as working platforms for construction equipment over very weak subgrades. Other paving materials cannot be placed directly on a weak subgrade as they break up during compaction. Generally, either the subgrade is cut into and backfilled with a large amount of granular material or a thick berm is constructed using granular materials. Excavation and granular material costs may be saved by using sand–asphalt–sulfur mixes.

Full depth sand–asphalt–sulfur pavement structures should reduce the depth of frost penetration into the subgrade in low temperature regions and thus reduce frost damage to pavements. The coefficient of thermal conductivity of these mixes is approximately one third the value for asphalt concrete.

Leveling Courses. Sand–asphalt–sulfur mixes may be used as thin leveling courses over deformed or worn pavement surfaces. The mix may be placed with a bituminous finisher, and a smooth finished surface is attained as the mix is not rolled and differential compaction is not a factor.

The wheelpath depressions of rutted and worn surfaces may be filled with mix, with the finisher screed riding on the bumps. Sand–asphalt–sulfur mixes can be feathered out to the thickness of the maximum-sized particle. The normal city street overlaying practice requiring heater-planing of the bumps and replacing of a full width overlay need not be followed.

Pavement Patching. Sand–asphalt–sulfur mixes are particularly suitable as pavement patching materials because the mixes may be poured into place and leveled. There is no problem of overfilling or underfilling the cavity, and the roadway may be opened to traffic soon after the mix cools. The process is amenable to replacing broken pavement surfaces, filling tops of cuts and trenches in city streets, repairing spalled portland cement concrete surfaces and joints, leveling road surfaces at railway crossings, etc.

The patching mix may be processed in an asphalt hot-mix plant and transported to the various patching sites in a heated vessel. Alternatively, the patching mix may be produced by remelting previously made hardened mix in a mobile heated vessel equipped with an agitator. The latter process is particularly adaptable for use during the winter in areas where hot-mix plants are not normally operational.

Bleeding Surface Overlay. Flushing and bleeding asphalt pavement surfaces may be overlaid directly with sand–asphalt–sulfur mixes without heater-planing the suface. Figure 4 indicates that the Asphalt Institute (*10*) recommended minimum Marshall stability requirements of 500 lb (2225N) and 750 lb (3335N) for medium and heavy traffic categories, respectively, may be satisfied even with mix formulations with high asphalt contents. Thus, these mixes may be used to overlay bleeding surfaces provided that the decrease in stability, associated with the upward migration of bleeding asphalt, does not drop below the recommended minimum requirements.

High Bearing Capacity Mixes. Conventional asphalt concrete mixes often deform under high loads over long loading periods. Thus conventional asphalt concrete is unsuitable for applications such as heavy duty industrial floors and container ports. The creep properties of such mixes may be improved considerably by adding sulfur to the mix.

Because of the increased workability of these mixes, careful mix handling is essential to avoid segregating the coarse aggregate from the finer mix material. Some consolidation effort is necessary for placing the mix so that the particles are oriented in the most dense configuration.

Porous Mixes. Specific aggregate–asphalt–sulfur formulations in which sulfur adds to the mix strength may be used for porous constructions through which water may drain readily. Such mixes permit, for example, the design of level tennis courts and may be used as porous drainage layers in hydraulic structures. These mixes also show potential as open-textured pavement surfaces. The high surface porosity may alleviate hydroplaning and is expected to exhibit good skid resistance under wet conditions and good anti-splash properties.

Sulfur-Extended Mastic Mixes. The use of sulfur in mastic or Guss-asphalt mix type formulations extends the use of this material. For example, softer asphalt grades may be used, and the mixes may be placed at lower temperatures. Using sulfur would permit variation in mix workability and control of the mechanical properties of the mixes and would extend the variety of aggregates that may be used.

The sulfur-extended mastics are high quality materials. They may be used in road surfacing and maintenance, railway beds, and as waterproofing materials for bridge decks and in roofing.

Hydraulic Applications. Since sand–asphalt–sulfur mix formulations can be highly impervious, they may be used in hydraulic applications. Because the material can be cast in place without compaction, it is ideally suited for placement on slopes where conventional materials are difficult to compact such as for lining storage reservoirs, sewage lagoons, and ditches which are susceptibel to erosion.

Pourable sand–asphalt–sulfur mix formulations may not be placed on exceedingly steep slopes as they would flow down the slope. In this case, stiffer aggregate–asphalt–sulfur mix formulations may be used.

Castings. Pourable sand–asphalt–sulfur mixes may be cast in various shaped molds. For example, the mix may be molded in the configuration of the New Jersey rigid median barrier (11). Stiffer mixes may be extruded with an asphalt curbing machine to form a high stability curb. A small slip-form paver may be adapted for casting sidewalks in a similar fashion.

Observations and Conclusions

Some specific conclusions were made from the laboratory and field program:

1. The liquid sulfur conforms to the shape of the voids in the mix and after cooling acts predominantly as a solid filler in the mix, formed *in situ*.

2. The workability of a mix, above the solidification point of sulfur, increases with increasing volume of fluids (asphalt plus sulfur). Properly designed mixes may be placed without roller compaction.

3. The workability of the mix is susceptible to temperature changes, and mix temperature control is important in handling, transporting, and placing the mix and during test specimen preparation and mix testing.

4. One-sized sands may be used to produce high stability mixes.

5. An optimum sulfur content exists for optimum fatigue behavior at any asphalt level in the mix.

6. Sand–asphalt–sulfur mixes exhibit a high degree of impermeability at higher air voids contents than conventional asphalt mixes.

7. In designing sand–asphalt–sulfur mixes, a number of mix properties such as workability, stability, fatigue behavior, and mix impermeability should be considered in establishing the optimum mix formulation for its intended use.

8. A number of test pavements have been constructed to evaluate mix durability and field performance variables under in-service conditions. While the results are promising, a longer test period is necessary before conclusive results are obtained.

By incorporating sulfur in asphalt mixes, high quality paving materials can be manufactured using inexpensive, poorly graded sands. These sand–asphalt–sulfur mixes may be used to construct road bases and surfaces, build platforms over weak subgrades, curbing, sidewalks, and castings of various shapes. In road maintenance, the mixes may be used to level courses, to overlay a bleeding pavement surface, and as a pourable patching material.

The use of sulfur in asphalt mixes permits the design of impervious materials for concrete bridge deck overlays and for use in hydraulic applications. Conversely, the mixes may be designed to be porous for drainage layers and anti-splash roadway surfacings with good skid resistance. Other specialty mixes which have been developed are high bearing capacity mixes and sulfur-extended mastic mixes that may be placed at lower temperatures than conventional mastics. Material application problems remain to be solved for several of these uses.

Literature Cited

1. Blue, D. D., Sullivan, T. A., "U.S. Bureau of Mines Research Program on New Uses for Sulphur," *Joint Chem. Eng. Conf.*, **4th,** Vancouver, Sept. 1973.
2. Deme, I., "Basic Properties of Sand-Asphalt-Sulphur Mixes," *Inter. Road Fed. World Meetg.*, **7th,** Munich, Oct. 1973.
3. Deme, I., "The Use of Sulphur in Asphalt Paving Mixes," *Joint Chem. Eng. Conf.*, **4th,** Vancouver, Sept. 1973.
4. Deme, I., "Processing of Sand-Asphalt-Sulphur Mixes," *Annu. Meetg. Ass. Asphalt Paving Technol.*, Williamsburg, Va., Feb. 1974.
5. Hammond, R., Deme, I., McManus, D., "The Use of Sand–Asphalt–Sulphur Mixes for Road Base and Surface Applications," *Proc. Can. Tech. Asphalt Ass.* (Nov. 1971) **16.**
6. American Society for Testing and Materials, "Concrete and Mineral Aggregates," Part 10, 1972.
7. American Society for Testing and Materials, "Bituminous Materials: Soils: Resistance to Plastic Flow of Bituminous Mixtures Using Marshall Apparatus," 1972, 493.

8. Heukelom, W., Klomp, A. J. G., "Road Design and Dynamic Loading," *Proc. Ass. Asphalt Paving Technol.* (1964) **33**.
9. Davies, J. R., Walker, R. N., "An Investigation on the Permeability of Asphalt Mixes," *Ontario Ministry Trans. Commun. Research Rep.* (1969) **145**.
10. Asphalt Institute, "Mix Design Methods for Asphalt Concrete," Manual Series No. 2 (MS-2), May 1963.
11. Smith, P., "Development of a Three Cable Guide Rail System and Other Guide Rail Tests, 1967-1968," Dept. of Highways of Ontario, Report No. **RR157**, Mar. 1970.

RECEIVED May 1, 1974

7

Beneficial Use of Sulfur in Sulfur–Asphalt Pavements

D. SAYLAK and B. M. GALLAWAY

Texas Transportation Institute, Texas A&M University,
College Station, Tex. 77843

H. AHMAD

Civil Engineering Department, Texas A&M University,
College Station, Tex. 77843

> *The influence of both material and process variables on mechanical properties and a preliminary comparison of the engineering characteristics of sand–asphalt–sulfur with standard asphaltic concrete pavements are evaluated. Material variables include sand:asphalt:sulfur ratios, asphalt type, and sand type and grading. Process variables include mixing temperature, mixing time, cure time, and compaction. Engineering properties such as stability, mixture density, strength, air void content, and thermal behavior are discussed. Concurrent with the laboratory evaluation program, odors and emissions were investigated by using evolved gas analysis techniques. Preliminary structural analysis of sulfur–asphalt mixtures were compared with conventional asphaltic concrete, using layered, elastic theory.*

In 1963 Shell Canada Limited initiated a laboratory program which eventually led to the development of a sand–asphalt–sulfur (S–A–S) paving material called Thermopave. This study revealed that cast S–A–S mixes with excellent engineering properties could be prepared by using locally available aggregates normally considered unacceptable for pavement construction. This concept was also investigated, to a lesser extent, by Shell Oil Co. and Shell Development Co.

In 1964, full-scale test sections were laid in Oakville, Ontario and St. Boniface, Manitoba, using conventional asphaltic concrete paving procedures. During these trials, rolling procedures routinely used on asphaltic concrete produced unacceptable pavements whereas an unrolled test section performed satisfactorily. It was later shown that conventional rolling of a mix behind a paver caused a breakdown in the macrostructural bonding (formation) between the constituents as the mix cooled.

In 1970, another field trial was conducted in Richmond, British Columbia where the Thermopave was essentially cast between wooden forms without densification or compaction. The successful results achieved on this project indicated that unrolled S–A–S mixes had potential as a paving material, both for surfacing and as a pavement base over a weak subgrade (1).

Interest in the United States to exploit this development stems from a desire to find efficient means of using the large quantities of sulfur which are expected to be produced by pollution control processes. To this end a continuing study under the joint sponsorship of the Bureau of Mines and the Sulphur Institute was initiated on May 1, 1973 to introduce the techniques now under development in Canada to researchers and the paving industry of the United States. The following discussion presents the results of this study.

Preparation of Sand–Asphalt–Sulfur Mixes

Sand–asphalt–sulfur mixes are prepared in the laboratory using two separate wet mix cycles. First, asphalt and sand preheated to 300°F are mixed for about 30 sec to coat the sand particles with asphalt. Sulfur, also at 300°F, is then added to the hot sand–asphalt and mixed for about 30 sec to achieve a uniform dispersion of the sulfur throughout the mix.

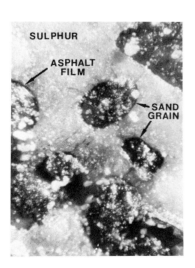

Figure 1. Photomicrograph of S–A–S matrix showing the mechemical interlock of sand particles provided by the sulfur (48×)

The molten sulfur, in addition to increasing the workability of the mix, normally fills the voids between the asphalt-coated sand particles. As the mixture cools below about 245°F, the sulfur solidifies, creating a mechanical interlock between the sand particles giving the mix a relatively high degree of stability. This interlocking effect is shown in Figure 1, which is a photomicrograph of a mix matrix, illustrating how perfectly the sulfur conforms to the geometry of the voids.

The S–A–S mixtures should be prepared and mixed while the materials are between about 260 and 320°F. The former represents the melting (solidification) point plus a tolerance to avoid sulfur structuring effects, and the latter is the temperature above which sulfur undergoes an abrupt and very large increase in viscosity as shown in Figure 2. Although these viscosity changes are perfectly reversible (2), they do adversely affect the workability of the mix above about 325°F. However, as will be shown later, acceptable mixes were prepared at sulfur temperatures

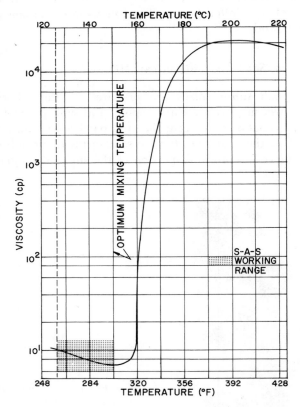

Figure 2. Viscosity–temperature curve for liquid sulfur

Table I. Physical Properties of Asphalts

Physical Properties	Asphalt I	Asphalt II
Specific gravity (77°F)	1.004	1.000
Pentration (77°F)	106	195
Viscosity, stokes (140°F)	1392	417
Viscosity, stokes (275°F)	3.2	1.5

near 380°F. At mixing temperatures below 240°F, the sulfur acts as a filler, giving the material the characteristics of a high filler content mix.

Materials

Elemental sulfur (*i.e.*, sulfur in the free state) of commercial grade (99.8± purity) was used throughout this investigation. Most of the mixtures were prepared with an asphalt conforming to 1972 Texas Highway Department Standard Specifications, Item 300 (*3*). Shell claims that standard penetration grade asphalts ranging from 40–300 may be suitably used in these mixes (*2*). The asphalt cement used in this study had a standard penetration of 106. As a check, additional verification tests were performed on some select compositions, using a 200 penetration asphalt. The physical properties of the two asphalts are given in Table I.

A particularly advantageous feature of sulfur–asphalt composites is that pavements prepared from heretofore unsuitable aggregates such as poorly graded sands usually have properties comparable or superior to pavements constructed with conventional asphaltic concrete mixes using high quality well graded aggregates. Therefore, four types of aggregates were tested: a poorly graded beach sand (Sand I) which was obtained from the Texas Gulf coast area, a more densely graded concrete sand (Sand II), a dense grade crushed limestone, and a rounded gravel. The physical properties of these aggregates shown in Table II reflect the relatively high voids content in the mineral aggregate (VMA) of the sands as compared with the crushed limestone and gravel.

Table II. Physical Properties of Aggregates

Designation	Aggregate Type	Specific Gravity	Voids in Mineral Aggregates (%)	Unit Wt (lb/ft^3)
Sand I	blow sand	2.65	37.6	103
Sand II	concrete sand	2.66	33.1	111
Limestone	crushed limestone	2.65	17.8	136
Gravel	rounded gravel	2.65	14.6	148

Laboratory Evaluations

Mix and material properties tests on a wide variety of S–A–S mixtures were performed using the aggregate and asphalt types discussed above. The specific mixture ratios evaluated ranged from 2:1 to 5:1 wt % sulfur to asphalt. The maximum amount of sulfur used in any mixture was 20 wt %. For comparison purposes, sand–asphalt (0% sulfur) and sand–sulfur (0% asphalt) mixes were also evaluated.

Test Methods and Evaluation Criteria. Specific tests which were performed include:

1. Marshall stability—ASTM D 1559-71
2. Percent air voids—Asphalt Institute method (4)
3. Unit weight—Texas Highway Department test method—Tex. 207-F (Rev. 1 Jan. 72)
4. Hveem stability—ASTM D 1560-65
5. Splitting tensile test—ASTM C 496-71
6. Thermal properties
    (a) Thermal conductivity—ASTM D 2214-70
    (b) Specific heat—ASTM C 351-61
    (c) Thermal expansion—ASTM D 696-70
7. Freeze–thaw—ASTM C 216-60, ASTM C 666-71

A more detailed listing of these ASTM methods is given at the end of this paper. The criteria suggested by the Asphalt Institute (5) and Shell (1, 2) for asphaltic concrete (A/C) under heavy, moderate, and light traffic conditions were used to evaluate these test results. Table III lists recommended test values (5) for conventional A/C mixes deemed necessary to withstand light, medium, or heavy traffic.

The criteria for maximum allowable air voids content in the final pavement was taken at 15% with a unit weight of about 125 lb/cu ft as established by Shell (2). Although many different mixture ratios were tested, the major comparisons were made with the Shell Thermopave mixture design, i.e., 80.5% sand–6% asphalt–13.5% sulfur by weight. Under different situations other designs may be equally attractive both technically and economically.

Sample Preparation. Standard Marshall molds (ASTM D-1559-71) were used to prepare samples for Tests 1–5. The aggregate, asphalt,

Table III. Suggested Test Values for Various Traffic Loadings

Design Method	Heavy Traffic	Medium Traffic	Light Traffic
Marshall stability (lb) (min.)	750	500	500
Marshall flow (1/100 in.)	8–16	8–18	8–20
Hveem stability (%)	37	35	30
Air voids (%)	3–5	3–5	3–5

and sulfur were preheated to 300°F in a thermostatically controlled oven. The mold assemblies were also heated to this temperature, and a thin coat of mold-release agent was applied to their inner surfaces to prevent sticking as the S–A–S mixture cooled.

A preheated mixing bowl was filled with a measured amount of preheated sand. Then the required amount of hot asphalt was weighed into the bowl with the sand and mixed thoroughly with a mechanical mixer for a predetermined period. This was followed by a second wet mix cycle using the molten sulfur. Each batch of mix contained enough material (approximately 8 lb) to prepare three test specimens.

The mixture was next poured into the preheated molds and compacted by rodding 25 times with a hot spatula. It was found early in the program that this compaction method disturbed the solidifying process of the sulfur and adversely affected the engineering properties measured. This procedure was eventually discontinued in favor of giving one face of the samples two blows using a 10-lb hammer with an 18-in. fall. This is consistent with the sample preparation techniques currently being used by Shell and indicates the need to keep consolidation and densification activity to a minimum. All asphaltic concrete samples were prepared according to ASTM procedures using 75 blows to each face.

After cooling in air for at least 24 hrs, the samples were extracted from the molds and made ready for testing in accordance with the standardized test methods given above.

The samples for Tests 6 and 7 were prepared in the same manner except in different test configurations. These specimen geometries will be discussed in the sections of this paper dealing with these tests.

Discussion of Results

Material Variables. Figures 3–12 illustrate the effect of material variables on the properties generated in Tests 1–4. The asphalt cement used in these tests was AC-10 (106 penetration) and the aggregate was the poorly graded beach sand (Sand I). Comparative results using the concrete sand (Sand II) and a softer asphalt AC-5 (195 penetration) will be presented later.

Figures 3 and 4 reflect the relationship between Marshall stability and the percent by weight of asphalt and sulfur, respectively. All points shown represent the average of at least three tests. These figures indicate the Marshall stability to be primarily affected by the sulfur content with a tendency to higher values as the sulfur content increases and lower values as the asphalt content increases. This is attributed to the fact that the additional sulfur reduces the voids content, thus enhancing the stiffness of the material. The additional asphalt, on the other hand, also

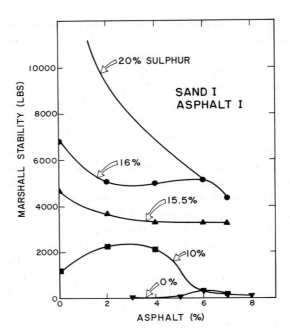

Figure 3. Marshall stability vs. asphalt content

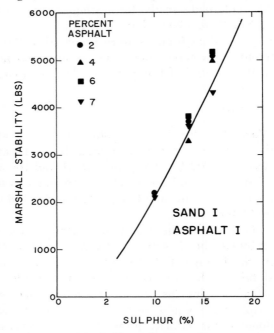

Figure 4. Marshall stability vs. sulfur content

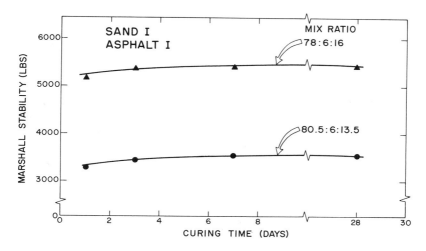

Figure 5. Effect of curing time on Marshall stability

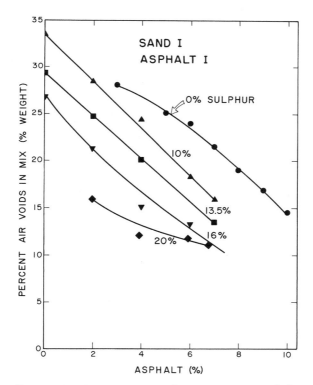

Figure 6. Percent air voids in mix vs. asphalt content

reduces voids content between the particles but produces a reduction in stiffness, thereby causing a decrease in stability. All mixtures except those with 0% sulfur and the 10% sulfur at asphalt contents in excess of 6% gave stability numbers acceptable for heavy traffic use. Figure 7 does indicate an ability to tailor the stability to satisfy individual needs and conditions.

Figure 5 shows virtually no post-cure change in the stability of two mixtures after 28 days. At the time of this writing, samples are in storage to extend this observation to 6 months. Both curves indicate that the stability improves slightly with age and essentially reaches full cure at approximately 6 days.

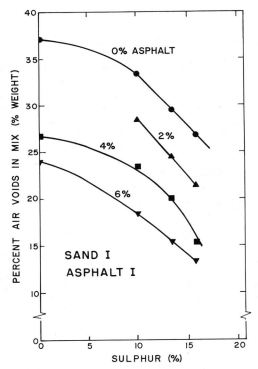

Figure 7. Percent air voids in mix vs. sulfur content

As discussed above, Figures 6 and 7 reaffirm that air voids are reduced as both asphalt and sulfur contents are increased. The Thermopave mixture ratio (S:A:S = 80.5:6:13.5) satisfies the 15% maximum allowable value prescribed by Shell (2). Figure 8 shows that the unit weight of the mix increases to some maximum value with asphalt as well as sulfur content. This, again, is caused by the physical properties of the

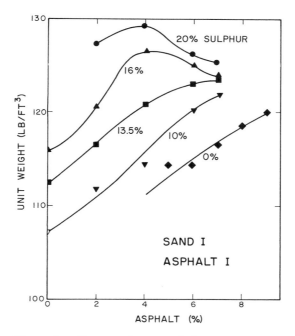

Figure 8. Unit weight vs. asphalt content

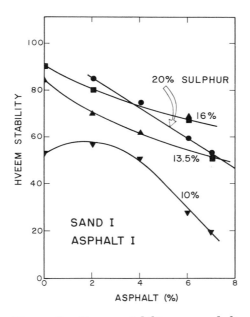

Figure 9. Hveem stability vs. asphalt content

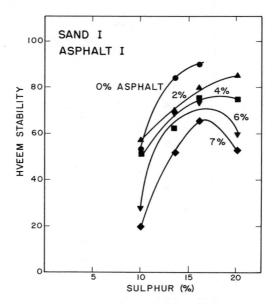

Figure 10. Hveem stability vs. sulfur content

aggregate and the void-filling action of the sulfur which results in density increase. The results indicate that the suggested value of 125 lb/ft³ (2 g/cc) can be achieved by using Sand I mixture designs with at least 13.5% sulfur. There is, however, no magic in the 125 lb/ft³ value.

Figures 9 and 10 show that Hveem stability will increase with sulfur content and decrease with increasing asphalt content. This is in agreement with the results observed in the Marshall tests. With the exception of 10% sulfur with asphalt contents greater than 5.5%, all mixes investigated exceeded the suggested test limits of 37 for heavy traffic shown in Table III.

Table IV represents a composite of the properties generated in Tests 1–4. In addition, the properties obtained using the more densely graded concrete sand (Sand II) are also given for comparison. Except for the 10% sulfur mixtures, the Marshall stabilities of the mixes using beach sand appear to be equal to or greater than those using the more expensive aggregate. This tends to become even more pronounced as sulfur content is increased. The Hveem stability was consistently higher for the beach sand mixes, indicating that the poorer graded aggregate was more desirable for S–A–S pavements. This is contrary to that expected and is believed to result from the effect of the stress concentrations in the crystallized sulfur produced by the higher angularity of the surface of Sand II. As will be shown later, the splitting tensile tests (Test 5) also indicated this superior mechanical behavior of pavements using Sand I.

The lower air void contents of the concrete sand are reflected in the higher unit weight and lower air void content of the mix, both of which increase with the amount of asphalt and sulfur present. No significant difference in the Marshall flow was revealed in using the two aggregates.

Table IV. Sulfur–Asphalt Mix Properties With Sands I and II (Asphalt I)

Asphalt (wt %)	Sulfur (wt %)	Marshall Stability (lb)		Marshall Flow (0.01 in.)		Air Voids (%)		Unit Wt (lb/ft³)		Hveem Stability	
		I	II	I	II	I	II	I	II	I	II
3	0	0	—	—	—	28.4	—	113	—	—	—
5	0	50	—	30	—	25.2	—	115	—	—	—
6	0	300	—	25	—	23.9	—	115	—	—	—
7	0	110	—	20	—	21.5	—	117	—	—	—
8	0	90	—	27	—	19.0	—	119	—	—	—
0	10	1250	—	13	—	33.4	—	107	—	53	—
2	10	2220	1980	6	6	28.3	21.2	112	123	57	48
4	10	2100	2110	5	6	24.3	16.2	115	127	51	47
6	10	300	1230	8	6	18.2	11.6	120	130	28	36
7	10	90	1000	7	7	15.9	10.5	122	130	20	30
0	13.5	4610	6240	5	8	29.4	24.6	113	120	84	87
2	13.5	3690	3640	7	7	24.7	15.7	116	131	70	60
4	13.5	3300	3530	7	7	20.1	10.6	120	135	62	58
6	13.5	3280	3510	7	6	13.2	8.5	127	134	69	54
7	13.5	3220	1510	7	8	13.5	6.9	124	134	52	34
0	16	6810	9150	3	8	26.8	20.6	116	126	90	90
2	16	5040	5170	7	7	21.4	12.4	121	135	80	65
4	16	4990	3800	6	8	15.1	8.9	127	136	75	60
6	16	5130	3500	7	8	13.4	7.9	125	134	68	55
7	16	4330	1940	7	8	13.0	6.6	124	134	51	38
0	20	—	19070	—	8	—	15.8	—	132	—	95
2	20	7770	8150	8	8	16.0	8.3	128	140	85	70
4	20	8560	4750	9	9	12.2	8.2	130	136	75	63
6	20	5190	4100	9	8	11.9	6.9	127	134	60	58
7	20	4520	3170	8	8	11.2	6.6	126	133	53	50

The effects of penetration grade (*i.e.*, relative hardness or consistency of asphalt) on the stability and air void content of S–A–S mixtures with 13.5% sulfur are shown in Figures 11 and 12, respectively. The penetration grades of Asphalt I and Asphalt II are 106 and 195, respectively, at 77°F. Tests results shown in Figure 11 indicate the Marshall stability of the mixes with Asphalt I to be greater than those using Asphalt II because of the relative softness of the higher penetration grade asphalt. Cured mixes with Asphalt II were more flexible and softer than those with Asphalt I. The air voids content did not appear to be significantly affected by the penetration grade of the asphalt. Acceptable

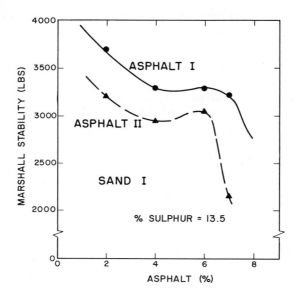

Figure 11. Effect of asphalt with different standard penetration grade on Marshall stability

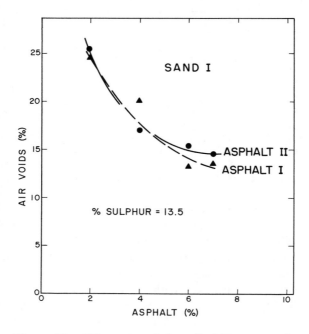

Figure 12. Effect of asphalt with different standard penetration grade on percent air voids

values of Marshall stability and air voids content of the mixes were obtained with either grade of asphalt as claimed by Shell (2).

Asphaltic Concrete. Nine asphaltic concrete mixtures prepared by conventional techniques using the crushed limestone and rounded gravel aggregates were used as a basis for comparison with S–A–S mixtures. The Marshall stability and flow, unit weight, percent air voids, and Hveem stability values for these mixtures with 75 compaction blows are shown in Table V. These data reflect the superior properties inherent

Table V. Asphaltic Concrete Mix Properties

Asphalt Content (%)	Marshall Stability (lb)	Marshall Flow (0.01 in.)	Unit Weight (pcf)	Air Voids (%)	Hveem Stability
Gravel Asphalt Concrete—75 Blows Compaction					
2.5	1628	7	144	9.4	41
3.0	1855	8	146	7.4	38
3.5	1793	8	148	5.2	36
4.0	1692	9	151	3.1	32
4.5	1610	12	152	1.6	20
5.0	1283	15	151	1.2	3
Crushed Limestone Asphaltic Concrete—75 Blows Compaction					
6.0	5287	11	142	6.0	48
6.5	4983	18	145	3.7	49
7.0	5140	15	145	2.9	49

6.5% Asphalt Mix Design Summary

	% Asphalt by Wt of Mix
Maximum density	6.7
Maximum Marshall stability	5.9
4% Air voids	6.0
Av.	6.2

Design Asphalt Content
6.2 wt % of mix
6.5 wt % of aggregate

in the mixes using the more expensive crushed limestone. The data represented in Figures 3–12 would indicate that, with the exception of unit weight and air voids, the properties of S–A–S mixtures can be tailored to provide the same range of values as those given in Table V. At higher sulfur contents, the stabilities of S–A–S mixtures exceeded those of the A/C with the crushed limestone. Because of the much higher VMA in Sands I and II, the air voids and unit weights of the A/C mixture could not be achieved by the S–A–S mixes. Unless otherwise noted, comparisons with other test results discussed in this report were made with the 6.5% asphalt mix, whose design summary is also given in Table V.

Effect of Process Variables

Mixing Time. The effects of mixing time on the Marshall stability and air voids content of S–A–S mixes are shown in Figure 13. The mixing time in seconds as shown on the abscissa of each graph is for the second wet mix process. For example, a mixing time of 20 sec indicates that sulfur was added to the sand–asphalt and mixed for 20 sec. Figure 13 indicates that the Marshall stability goes through a maximum or optimum value for a mixing period of 30 sec. The minimum value of air voids content was also obtained at 30 sec. This is almost twice the time reported by Shell (1) in their pug mill plant operation. At least 30 sec were required in the first wet mix to coat the surfaces of the sand particles with asphalt. These data also indicate that if excessive mixing times are used, both stability and air voids are adversely affected. This is attributed to the fact that the cooling rate of the mixture increases with mixing and disturbs the natural crystallization process of the sulfur within the voids between the asphalt-coated aggregate, thus destroying the mechanical interlock provided by the sulfur.

The optimum mixing time will depend on the amount of material to be mixed and the size of the mixer. The relationships shown in Figure 13 were based on an 8-lb batch. The materials were mixed in a 2-gal. Hobart mixer.

Compaction. Once the mixture was poured into the mold, it was compacted by delivering a number of blows with a 10 lb weight to one face of the sample. In conventional A/C systems, this reduces air void content and increases unit weight of the cured material. The influences

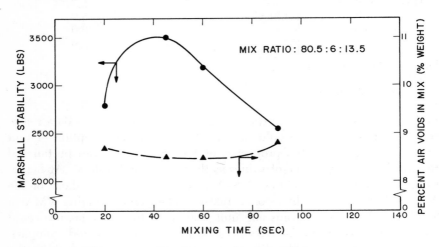

Figure 13. Effect of mixing time on Marshall stability and percent air voids

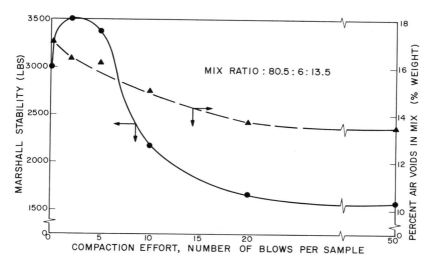

Figure 14. Effect of compaction on Marshall stability and percent air voids

of this compaction, as revealed by the effect on the Marshall stability and air voids content of a S–A–S mix, are shown in Figure 14. The specimen, without any compaction at all, has a Marshall stability value of 3000, and an optimum value is obtained by applying two blows to the sample. A compaction of more than two blows decreases the stability, apparently by destroying the mechanical interlock provided by the sulfur as it solidifies. Figure 14 also indicates that, as the number of compaction blows is increased, the air voids content in the mix decreases, reflecting higher degree of consolidation and densification produced. With the exception of the samples used in these particular tests, all specimens used throughout the program were given two blows compaction.

In addition to the problems associated with excessive mixing and compaction, a series of tests was designed to illustrate the extent to which the structural integrity of the S–A–S material would be affected by a loss of process control. The two operational problems simulated were: material handling below the solidification temperature of the sulfur (240° F) and a loss of temperature control at the hot-mix plant, causing a temperature rise above 320°F.

One batch of S–A–S material with 13.5% sulfur was prepared in the normal procedure and allowed to cool to 200°F prior to placing into three Marshall molds. This was followed by two blows to one side of each sample while in the cooled (200°F) condition. A second batch was prepared, cast, and compacted at 380°F. The latter was well above the polymerization point of the sulfur. Although the workability decreased slightly, the added stiffness of the mix did not hinder the preparation of the Marshall specimens.

The resulting Marshall stabilities of these specimens are compared with that of a normally prepared mix in Figure 15. Whereas the speci-

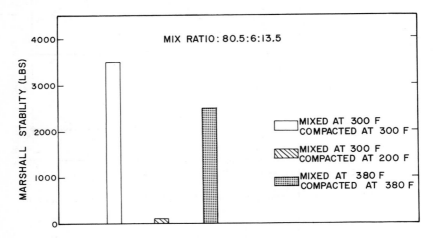

Figure 15. Effect of mixing and compaction temperatures on Marshall stability

mens prepared at the higher temperatures still maintained a respectable stability of 2500 lb, the stability of samples mixed at 300°F and compacted at 200°F was reduced to below 100 lb. Examination of the specimens following testing revealed a nonuniform dispersion of the sulfur across the thickness which would make the mix totally unacceptable. These tests indicate the importance of maintaining good temperature control while preparing, placing, and compacting the mix.

Splitting Tensile Test. Results of the splitting tensile tests for various S–A–S mix designs are shown in Table VI. These are compared with the conventional asphaltic concrete mix design shown in Table V.

In general, the splitting tensile strength of a mix increases with sulfur content and decreases with increasing asphalt content. Beach sand (Sand I) gave higher strengths than the concrete sand (Sand II). One reason for this effect could be that the higher degree of particle angularity in the concrete sand particles created stress concentrations in the sulfur crystals which produced premature failure in the specimen when the load was applied. This is also consistent with the results of the Marshall and Hveem tests discussed earlier. The degree of superiority of Sand I over Sand II varies from test to test, but the consistently higher strengths of the beach sand provide an added measure of confidence in the use of this aggregate.

The results tabulated in Table VI also give a first indication of the temperature and rate dependence of S–A–S materials. The effect of specimen temperature on strength was determined by conducting the test at three temperatures: 20, 73, and 135°F. The strengths decreased with temperature with the greatest change occurring between 20 and 73°F. The strength changed by a factor of 6 for S–A–S mixes and 12 for

Table VI. Effect of Temperature and Deformation Rate on Splitting Tensile Strength of Various S–A–S Mix Ratios

Test Temp. (°F)	Deformation Rate (in./in.)	Asphalt (wt %)	Sulfur (wt %)	Splitting Tensile Strength (psi)		
				Sand I	Sand II	A/C
20	2	6.0	13.5	309	303	—
		6.0	16.0	312	306	—
		6.0	20.0	312	305	—
		6.2	—	—	—	690
73	2	6.0	13.5	116	87	—
		6.0	16.0	127	79	—
		6.0	20.0	123	87	—
		6.2	—	—	—	156
135	2	6.0	13.5	28	17	—
		6.0	16.0	37	18	—
		6.0	20.0	43	23	—
		6.2	—	—	—	54
20	0.2	6.0	13.5	302	267	—
		6.0	16.0	327	267	—
		6.0	20.0	315	270	—
		6.2	—	—	—	685
73	0.2	6.0	13.5	62	42	—
		6.0	16.0	79	42	—
		6.0	20.0	80	51	—
		6.2	—	—	—	84
135	0.2	6.0	13.5	13	8	—
		6.0	16.0	18	8	—
		6.0	20.0	26	11	—
		6.2	—	—	—	21
20	0.02	6.0	13.5	307	277	—
		6.0	16.0	292	283	—
		6.0	20.0	323	287	—
		6.2	—	—	—	685
73	0.02	6.0	13.5	38	23	—
		6.0	16.0	50	21	—
		6.0	20.0	54	26	—
		6.2	—	—	—	56
135	0.02	6.0	13.5	7	3	—
		6.0	16.0	9	4	—
		6.0	20.0	14	7	—
		6.2	—	—	—	17

A/C at the lower deformation rates over this temperature range. The strengths of the asphaltic concrete samples were nearly the same as those for the S–A–S mixed with 20% sulfur at 73 and 135°F but peaked at 690 psi at 20°F.

The effect of deformation rate on splitting tensile strengths was determined using three speeds (2, 0.2, and 0.02 in./min). The results shown in Table VI indicate that the strength is directly related to the loading rates, reaching a maximum of about 300 psi at 20°F with each rate. The 0.02 in./min deformation rate showed the greatest change in strength with the 20% S–A–S mixture and the A/C mix, increasing by 20 and 40%, respectively.

Since reliable strain data could not be obtained the stiffnesses in these mixes could not be calculated. A cursory examination of the data indicates that the stiffness of S–A–S is higher than A/C at very low rates. As the loading rates increase, the difference in rate dependence of the two materials produces comparatively greater changes in stiffness in the A/C than in the S–A–S system. This is consistent with the results of tests conducted by Shell which indicate that the stiffness of S–A–S is about 20 times greater than A/C under constant-load creep at 70°F (2). Other data (1) shows the dynamic stiffness of S–A–S and A/C to be 660,000 and 900,000 psi, respectively, at 50 Hz. It would appear that at light-to-moderate loading rates the S–A–S mixtures would have higher stiffness. This estimate was used in setting up the conditions for the preliminary theoretical analysis to be discussed later in this report.

The results of these tests are inconclusive in a quantitative sense, but do demonstrate a higher degree of viscoelastic behavior in the asphaltic concrete than in the S–A–S systems. This is to be expected with crystalline sulfur in the matrix. Any viscoelastic behavior experienced in the S–A–S mixes was primarily attributed to the asphalt.

Thermal Properties

Thermal Conductivity. The results of the thermal conductivity, k, tests are given in Table VII for four different mixture ratios representing S/A ratios of 1.3 to 10. The data indicate that additional sulfur had little effect on the thermal conductivity, which averaged 11.7×10^{-4} cal-cm/cm^2-sec-°C ($3.40 \times$ Btu-in./ft^2-hr-°F). A comparison with the value obtained for the A/C system of 15.77 cal-cm/cm^2-sec-°C (4.57 Btu-in./ft^2-hr-°F) would indicate that the thermal conductivity is about 25% less for S–A–S than for A/C. This is attributed to the higher air void contents in the former which add to the insulative characteristics of the material.

The TTI-measured thermal conductivity value of 17.2×10^{-4} cal-cm/cm^2-sec-°C (5.0 Btu-in./ft^2-hr-°F) is considerably less than that obtained by Shell (6). The Shell data were obtained by using the guarded hot plate method (ASTM C-177), which is more accurate and consistently generates higher k values than the apparatus used in this study. Therefore, the results generated in this phase of the program are at best qualitative but do indicate that some differences in thermal expansion between adjacent layers of S–A–S and A/C may result from the mismatch in their thermal conductivities.

For comparison purposes, the published k values of sand, asphalt, limestone, and sulfur are also given in Table VII. On the basis of the range of values for limestone, the experimental result for asphaltic concrete appears reasonable.

Table VII. Thermal Properties of S–A–S Mixtures vs. Mixture Ratio[a,b]

Mixture Ratio S–A–S	Thermal Conductivity (cal-cm × 10^{-4}/ cm^2-sec-°C)	Coefficient of Thermal Expansion (°C^{-1})	Specific Heat (cal/lb-°C)
	Experimental Values		
84.5:2:13.5	11.8	26.6 [25.0]	0.202
83:7:10	11.1	—	0.230
80.5:6:13.5	10.7	29.4 [29.0]	0.224
80.5:6:13.5	10.2	29.3 [29.0]	0.224
78:2:20	12.2	26.5 [26.8]	0.197
74:6:20	—	41.9 [42.6]	0.217

Other Materials	Thermal Conductivity (cal-cm × 10^{-4}/ cm^2-sec-°C)	Coefficient of Thermal Expansion (°C^{-1})	Specific Heat (cal/lb-°C)
	Published Values		
Asphalt	29.6 (7)	200 (8)	0.430 (7)
Asphaltic-concrete	15.7 (7)	21–25 (9)	0.18–0.22 (10)
Sulfur	6.0 (11)	64.1 (12)	0.137 (13)
Sand	8.6 (12)	5.9–12.5 (14)	0.201 (15)
Limestone	13.8–30.9 (13)	10.8–85.0 (14)	0.203 (13)

[a] Numbers in brackets are values obtained using the Rule of Mixture.
[b] Numbers in parentheses are references.

Thermal Expansion. Experimental results obtained from S–A–S mixtures and a conventional asphaltic concrete are also given in Table VII. Published data on asphalt cement, asphaltic concrete sulfur, sand, and limestone are also provided. The overall thermal expansion coefficient of the composite is derived from the combined effects of the individual ingredients in the mixture and the air voids present in the final material. Any combination which tends to decrease the air voids content

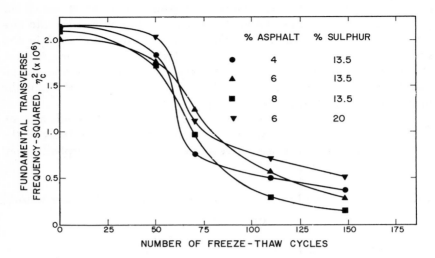

Figure 16. Fundamental transverse frequency-squared vs. number of freeze–thaw cycles

will produce an increase in the net thermal expansion. The values obtained for the S–A–S mixture ranged from 4 to 100% higher than the published values for asphaltic concrete.

The experimental data correlated quite well with those computed using the Rule of Mixtures. The computed values are given in parentheses in Table VII. The Thermopave S–A–S mixture (80.5:6:13.5) had a thermal expansion coefficient of 29.3×10^{-6} in./in.-°C which is about 30% higher than that of the A/C material used for comparison. This difference could have a significant effect on the stresses developed at the interfaces between adjacent layers of A/C and sulfur–asphalt mixtures. At this writing a more in-depth evaluation of the effects of the missmatch in thermal properties is in progress at the Texas Transportation Institute.

Specific Heat. The results of the specific heat tests are given in Table VII along with published values for some of the individual ingredients. Consistent with the Rule of Mixtures, the average values reflect the specific heat of the ingredient present in the largest quantity. For S–A–S mixtures the trend is established by the change in quantity of the asphalt. Of the two, the A/C system appears to be the slightly better heat sink.

Freeze–Thaw Test. A limited number of mixes were subjected to a series of freeze–thaw cycles in a chest-type freezer at a rate of 6–10 cycles/day. The fundamental transverse frequency at zero cycles, η, was measured at the beginning of each test. Additional values, η_c, were also measured at 50, 70, 110, and 150 cycles. The fundamental transverse frequency-squared *vs.* the number of freeze–thaw cycles is plotted in

Figure 16 and shows a rapid drop of η_c^2 values after 50 cycles. Since η_c^2 is directly related to the change in dynamic modulus, E_c, in accordance with:

$$\% \, E_c = \left(\frac{\eta_c}{\eta}\right)^2 \times 100\%,$$

This sudden drop indicates an adverse change in structural integrity of the mixtures after 50 cycles. Shortly after these initial tests were run, the freeze–thaw equipment began to malfunction. Therefore, it is not possible to make any definite statements, at this writing, regarding the effect of other mix variables on the freeze–thaw durability. Resistance to freezing and thawing was improved with mixes having higher sulfur contents. This was attributed to the reduced air voids associated with these mixes.

Evolved Gas Analysis

The odors and emissions evolved during the laboratory mixing and handling operations were evaluated by using a portable evolved gas analysis (EGA) technique. Under the influence of heat, sulfur has a condensing effect upon asphalt, forming gaseous hydrogen sulfide in accordance with the following reaction (16, 17):

$$C_xH_y + S \xrightarrow{\text{heat}} C_xH_{y-2} + H_2S \, (g)$$

Since the evolution of hydrogen sulfide could have a major bearing on the final acceptance of the S–A–S material in pavement construction, a study of the extent of hydrogen sulfide gas evolution encountered over a range of process variables was undertaken. A number of concentration levels and their associated environmental effects are shown in Table VIII.

Table VIII. Toxicity of Hydrogen Sulfide Gas

H_2S Concentration (ppm)	Environmental Impact
0.02	odor threshold value
5–10	suggested MAC
20	MAC (18)
70–150	slight symptoms after exposure of several hours
170–300	maximum concentration that can be inhaled for 1 hr without serious consequences
400–700	dangerous after continuous exposure of 0.5–1 hr
600	fatal with exposure greater than 0.5 hr

One accepted value of the maximum allowable concentration (MAC) of hydrogen sulfide to which a person can be exposed without injury to health is 20 ppm as set by the American Conference of Governmental Industrial Hygienists (ACGIH) (18). Although this level is not injurious to health, it does irritate the eyes. To prevent this irritation, a MAC of 5–10 ppm has been suggested (18). These two values were used to evaluate the relative toxicity of the hydrogen sulfide emissions in the preparation of the S–A–S mixes.

During the preparation of S–A–S samples, hydrogen sulfide gas evolution was noticed after adding sulfur and during compaction of the samples. A Metronics model 721 Rotorod gas sampler (19) was selected for use in these tests because it was portable, sensitive, and yet simple to operate.

During the mixing procedure, hydrogen sulfide was monitored by sampling the air 12–18 in. above the mix for a predetermined time. This distance was considered an "average working distance" of the technician preparing the mix. During each mix, care was taken to place the gas sampler in the same position relative to the mix. For this reason, concentrations of hydrogen sulfide detected by these tests are considered representative of the levels to which a worker would be exposed while preparing S–A–S samples in the laboratory.

Samplings were taken during the mixing and compaction phases of sample preparation. The results of these tests are given in Table IX and show that the concentrations of hydrogen sulfide evolved ranged from 0.1 to 2.75 ppm. These concentrations are well within the safe level for exposure to hydrogen sulfide as set forth in Table VIII.

The results of this study showed that the mean concentration of hydrogen sulfide evolved during preparation of S–A–S paving materials in the laboratory were well below the suggested maximum allowable concentration level. Peak concentrations higher than the maximum allowable levels were encountered. These, however, occur for such a short period of time that, in a well ventilated laboratory or remotely controlled commercial mixers, the preparation of S–A–S mixtures would not normally produce harmful amounts of hydrogen sulfide.

Table IX. Hydrogen Sulfide Evolution During Initial Samplings

Sulfur–Asphalt Ratio	Mean H_2S Concn (range, ppm)
2.25	0.54 (0.2 –1.8)
2.66	0.57 (0.5 –0.6)
3.33	0.54 (0.5 –0.6)
6.75	0.43 (0.3 –0.5)
8	0.17 (0.15–0.2)
10	0.35 (0.8 –0.1)

Preliminary Theoretical Analysis

General. The relative fatigue life of the four pavement sections shown in Figure 17 were evaluated using a layered elastic analysis. Both the loading conditions and the assumed elastic properties of the materials are given. The BISTRO layered elastic analysis computer program (20, 21) was used to determine the stresses and strains at various strategic locations in the pavement sections. These locations were directly under the load and at radial distances of 3 and 6 in. In Systems 2 and 4, stresses and strains were computed at the interface of the base material and the surface layer and the interface between the base material and subgrade. For Systems 1 and 3, stress and strains were determined only between the pavement layer and the subgrade. These locations are repre-

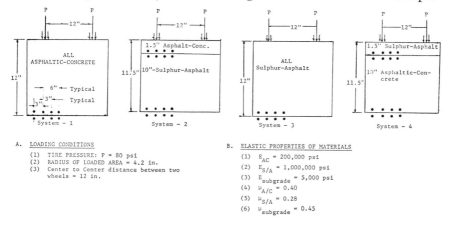

Figure 17. Pavement test sections used in fatigue life analysis

sented by dots on the pavement structure shown in Figure 17. In this preliminary analysis the following assumptions were made concerning the elastic properties of the sulfur–asphalt system:

1. The stiffness modulus of the sulfurized asphalt paving material can be five times greater than that of asphaltic concrete at intermediate rates of loading.

2. Poisson's ratio of S–A–S paving material is 0.28.

3. Shell design fatigue criteria (20) for asphaltic concrete and sulfurized asphalt paving material are identical.

Assumption 1 was made by interpolation from data reported by Shell Canada (1, 2). According to Shell Canada findings, the stiffness modulus of a S–A–S mix at 50°F and 50 Hz is about two-thirds of the Shell design charts modulus used for asphaltic concrete. At lower loading rates (creep rates) and 70°F the stiffness modulus of S–A–S mix is about 20 times larger than that of asphaltic paving mix.

Assumption 2 was made on the basis that the Poisson's ratio of the stiffer sulfurized asphalt mix should be appreciably lower than that of conventional asphaltic concrete. The value of Poisson's ratio, although required for the analysis, has a minimal effect on the stress and strain values calculated.

Assumption 3 was used because of the absence of an appropriate fatigue design criterion for the sulfur–asphalt pavement material. The fatigue life was evaluated on the ability of the four pavement systems to withstand an 18 kip axle load for 10^6 applications.

Table X. Results of Fatigue Life Analysis[a]

System	Pavement Section	Maximum Tensile Strain in Asphaltic Concrete or S–A–S Paving Layer ($\times\ 10^{-6}$ in./in.)	Fatigue Life of Paving Layer (passes of 18-Kip Axle)	Maximum Compressive Strain in the Subgrade ($\times\ 10^{-6}$ in./in.)
1	12-in. asphaltic concrete paving section	212	170,000	465
2	1.5-in. asphaltic concrete over 10-in. sulfur–asphalt base course	$\varepsilon_{A/C} = 69$ $\varepsilon_{S-A-S} = 73$	>1,000,000 >1,000,000	188
3	12-in. sulfur–asphalt paving section	62	>1,000,000	155
4	1.5-in. sulfur–asphalt surface course over 10-in. asphaltic concrete base course	$\varepsilon_{S-A-S} = 35$ $\varepsilon_{A/C} = 201$	>1,000,000 210,000	427

[a] Fatigue life of subgrade to achieve permanent deformation (passes of 18-kip axle) was > 1,000,000 in all systems.

Discussion of Results. The results of the relative fatigue life analysis of the four pavement sections are given in Table X. The analysis indicates that the 12-in. asphaltic concrete pavement (System 1) and the 1½-in. sulfur–asphalt over 10-in. asphaltic concrete base (System 4) were not able to withstand the 10^6 passes of an 18-kip axle load. For System 4, the 1½-in. S–A–S surfacing was adequate, but analysis indicated that the pavement would fail at the base of the 10-in. A/C layer. However, a comparison of the two systems indicates that the fatigue life of the asphaltic concrete pavement was moderately improved just by the presence of 1½-in. sulfur–asphalt surface.

The all-sulfur asphalt pavement (System 2) and System 3 with 1½-in. A/C surfacing over 10-in. S–A–S base were sufficiently adequate to satisfy the design criteria. For all pavement systems analyzed, subgrade failure was not predicted in 10^6 passes of an 18-kip axle load. Shell (*20*) criteria were used for subgrade failure criteria.

It should be re-emphasized that this analysis was only preliminary and based on assumed elastic properties of the sulfur–asphalt material. A more representative analysis using dynamic properties obtained from fatigue and resilient modulus tests is now in progress.

Conclusions and Recommendations

A series of standard pavement evaluation tests have been performed on a large number of mixtures comprised of different percent-by-weight ratios of sand, asphalt, and sulfur. The testing program performed in this task was designed to evaluate qualitatively the influence of both material and process variables on engineering properties. The following represent some of the conclusion reached:

1. S–A–S mixtures made with inexpensive, poorly graded sands has properties at least equal to or better than conventional asphaltic concrete.

2. For best workability and strength, processing should be accomplished at $250°F \leq T \leq 315°F$. Results have shown that mixtures can be prepared at temperatures as high as 395°F but at the sacrifice of a uniformly dispersed sulfur phase and excessive evolution of hydrogen sulfide.

3. No adverse trends in properties were indicated after 28 days of post-cure.

4. Engineering properties of 80.5:6:13.5 S–A–S mixtures prepared by TTI are in good agreement with Shell Thermopave material.

5. The design properties of a mix can be tailored to a predetermined stability, percent air voids, or unit weight by adjusting sulfur content.

6. Thermal properties of S–A–S appear to be relatively unaffected by amounts of sulfur and asphalt present over the range of S/A ratios tested.

7. S–A–S materials should be prepared with a minimum of compaction or densification. This could result in a significant economic advantage of S–A–S over A/C during construction.

8. Preliminary layered elastic analysis indicates that fatigue life of S–A–S at low-to-moderate rates of loading is better than asphaltic concrete and relatively equal at high rates.

9. Mean hydrogen sulfide concentrations measured during laboratory mixing and casting was below the maximum allowable concentration suggested by ACGIH.

10. Economics of using S–A–S as a pavement material depends to a large extent on availability of low-cost aggregate and the cost and accessibility of sulfur.

ASTM Test Methods Cited

ASTM C 215-60, "Fundamental Transverse, Longitudinal and Torsional Frequencies of Concrete Specimens," Part 10 (1970).
ASTM C 351-61, "Mean Specific Heat of Thermal Insulation," Part 14 (1967).
ASTM C 496-71, "Splitting Tensile Strength of Cylindrical Concrete Specimens," Part 10 (1973).
ASTM C 666-71, "Resistance of Concrete to Rapid Freezing and Thawing," Part 10 (1973).
ASTM D 696-70, "Coefficient of Thermal Expansion of Plastics," Part 27 (1971).
ASTM D 1559-72, "Resistance to Plastic Flow of Bituminous Mixtures Using the Marshall Apparatus," Part 11.
ASTM D 1560-65, "Resistance to Deformation and Cohesion of Bituminous Mixtures," Part 11 (1971).
ASTM D 2214-70, "Estimating the Thermal Conductivity of Leather with the Cenco-Fitch Apparatus," Part 15 (1971).

Literature Cited

1. Hammond, R., Deme, I., McManus, D., "The Use of Sand–Asphalt–Sulphur Mixes for Road Base and Surface Applications," *Ann. Meetg. Can. Tech. Asphalt Ass.*, Montreal, Nov. 1971.
2. U.S. Patent No. 3,738,853 (June 12, 1973).
3. Texas Highway Department, Standard Specifications for Construction of Highways, Streets, and Bridges, 1972.
4. "Mix Design Methods for Asphalt Concrete and Other Hot-Mix Types," "The Asphalt Institute Manual," Series No. 2 (**MS-2**), 3rd Ed., Oct. 1969.
5. "The Asphalt Institute Handbook," Manual Series No. 4 (**MS-4**) Mar. 1970.
6. Buxton, H. L., Shell Canada, Oakville Research Center, Toronto, Canada, private communication, 1974.
7. Barth, E. J., "Asphalt—Science and Technology," pp. 303-304, Gordon and Breach Science, 1962.
8. Pfeiffer, J. P., "The Properties of Asphaltic Bitumen," p. 91, Elsevier, 1950.
9. Monismith, C. L., Secor, G. A., Secor, K. E., "Temperature-Induced Stresses for Asphaltic Concrete," *Proc. Ass. Asphalt Paving Technol.* (1966) **35**.
10. Corlew, J. S., Dickson, P. F., "Methods For Calculating Temperature Profiles of Hot Mix Asphalt Concrete As Related to the Construction of Asphalt Pavements," *Proc. Ass. Asphalt Paving Technol.* (1968) **37**, 114.
11. "International Critical Tables of Numerical Data," Vol. 5, p. 217, McGraw-Hill, New York, 1929.

12. "Hudson's Engineers' Manual," 2nd Ed., pp. 314-317, John Wiley and Sons, Inc., New York, 1939.
13. "Handbook of Chemistry and Physics," 33rd Ed., p. 2041, CRC Press, Cleveland, 1963.
14. Roark, R. J., "General Engineering Handbook," 2nd Ed., p. 181, McGraw-Hill, New York, 1965.
15. Hadley, W. O., Hudson, W. R., Kennedy, T. W., "An Evaluation of Factors Affecting the Tensile Properties of Asphalt Treated Materials," Research Report **98-2**, Center for Highway Research, University of Texas, Austin, Mar. 1969.
16. Gamson, B. W., U.S. Patents **2,447,004, 2,447,005,** and **2,447,006,** Aug. 17, 1948.
17. Abraham, H., "Asphalts and Allied Substances," Vol. 1, 6th Ed., pp. 48–50, D. Van Nostrand, Princeton, Sept. 1960.
18. Elkins, H. B., "The Chemistry of Industrial Toxicology," pp. 95, 232, John Wiley and Sons, New York, 1950.
19. "Rotorod Gas Sampler for Hydrogen Sulfide," Metronics Technical Bulletin No. **9-72**, Metronics Associates, Inc., Palo Alto, Calif.
20. Izatt, J. O., Lettier, J. A., Taylor, C. A., "The Shell Group Methods for Thickness Design of Asphalt Pavements," *Ann. Meetg. Natl. Asphalt Paving Ass.,* San Juan, Puerto Rico, Jan. 7-13, 1967.
21. Koninklijke/Shell-Laboratorium, Amsterdam.

RECEIVED May 1, 1974

8

Sulfur/Asphalt Binders for Road Construction

C. GARRIGUES and P. VINCENT

Société Nationale des Pétroles d'Aquitaine, Tour Aquitaine,
92080 Paris La Defense, France

> *In a new process developed by SNPA, sulfur is dispersed in asphalt under patented conditions to form a sulfur/asphalt (S/A) binder. Conventional mix-plant and paving equipment is used to produce mixtures using this binder. The key advantage of this process is that sulfur replaces a significant portion of asphalt (30%). Total binder content of the mix, however, is similar to that normally used for asphalt alone. Preparation of the paving mixtures consumes less energy. Their application is easier and the compacting temperature is reduced. The improved pavement characteristics allow replacement of conventional aggregates by low quality aggregates or reduction of the thickness of the layers with graded aggregates.*

Disposal of increasingly larger amounts of sulfur recovered from coal, natural gas, and petroleum is expected in a near future as a result of air pollution abatement regulations. As a result, some years ago Société Nationale des Pétroles d'Aquitaine (SNPA) initiated a sulfur/asphalt research and development program to find a market for the elemental sulfur that will be produced if the current air quality standards are met.

At the same time, diminishing reserves of petroleum may make it desirable to reduce the consumption of petroleum products, including asphalt, where satisfactory substitutes can be found. Developments in sulfur/asphalt technology may permit partial replacement of increasingly scarce asphalt by increasingly abundant sulfur.

The process developed by SNPA involves a dispersion of sulfur in asphalt to form a sulfur/asphalt (S/A) binder. This S/A binder is then used to produce hot-mix paving mixtures with conventional mix-plant and paving equipment.

Although attempts to incorporate sulfur in asphalt were initiated many years ago (1, 2) and different authors investigated the sulfur/asphalt system (3, 4, 5, 6), the SNPA process produces very fine particles because of specific dispersion conditions. This important dispersion results in a stable high sulfur-content emulsion which has specific viscoelastic properties in a very large temperature range (especially low temperatures).

Preparation and Testing of Sulfur/Asphalt Binders

Sulfur dispersion in asphalt can be produced by various mechanical means. If the work is carried out over 120°C, sulfur emulsions in asphalt can be obtained since sulfur liquifies at 117°C.

Specific conditions used by SNPA allow production of sulfur particles having an average size of less than 5 μ. After 10 hrs, special stabilizing agents must be used.

If an emulsion of this type is allowed to cool in the open air, it should be checked to see that no emulsion breaking occurs (the sulfur droplets remain separate) but that sulfur enrichment at the lowest points is discernible. This sedimentation only occurs when asphalt is relatively fluid, as a result of the substantial difference in density between sulfur and asphalt and where the total sulfur content is greater than 20%.

In practice, by about 80°C, sedimentation has slowed down, and the emulsion can be held at 130°C for 10 hrs without marked apparent enrichment at the low points. If emulsions of this type are taken in the mass, they can be remelted without "breaking."

In all cases—hot storage or remelting—the homogeneity of the emulsion mass can be re-established by a simple pumping operation without any turbine.

Emulsions can be prepared with modified standard types of turbines. After appropriate adjustment, S/A binders of the desired quality are produced with an acceptable output/power consumption ratio. This laboratory has produced most of the binders described in the present paper, using the basic parts of a Moritz BF 50 turbine at 140°C for 5 mins, having first checked that this process incurred no risk of asphalt dehydrogenation.

The size of the sulfur particles is checked by microphotography ($\times 530$ magnification). Crystalline sulfur content is determined by differential enthalpic analysis, using a Perkin Elmer DSC 1 unit. Strictly controlled operating methods have been developed to avoid the phenomena of over-fusion. Results obtained are reproducible within $\pm 3\%$. Under these conditions, 10–20% by weight of sulfur added to asphalt is no longer present in the crystalline state. Results vary slightly according to the nature of the asphalt (origin and penetration) and the emul-

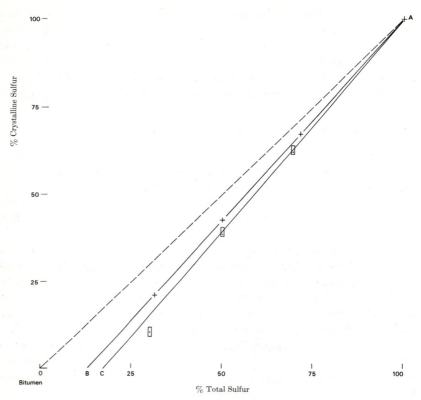

Figure 1. Solubility of sulfur. Crystalline sulfur in 80/100 bitumen. OA = forecast curve without solubility. BA = curve at 150°C (emulsification temp) during 5'. CA = curve at 180°C (emulsification temp) during 5'. OB = solubility after heating at 150°C ~ 14%. OC = solubility after heating at 180°C ~18%.

sification temperature, but are independent of sulfur content. The results for pen. 80/10 asphalt at two emulsification temperatures show 14% solubility by weight of elementary sulfur at 150°C and 18% solubility by weight at 180°C (Figure 1).

The difference in sulfur solubility between 150°C (no asphalt dehyrogenation reaction) and 180°C (effective asphalt dehydrogenation) is negligible. The solubilization phenomenon appears distinct from that of dehydrogenation.

Micrographic examination verifies that the addition of 10% sulfur to asphalt produces a monophasic product. The physical methods for determining these solubilities have been completed by chemical determinations of the sulfur content in each phase: total sulfur dispersed and sulfur content of each part of the bituminous phase (maltenes, asphaltenes, resins). Fine determination of the chemical state of dissolved sulfur is being performed, but it is already known that the results are different from those obtained with physical methods. As a matter of fact, solubilities appear to be more important. Physical determinations must be performed in a short time, and the chemical analyses show that the time is a very important factor in the dissolution mechanism.

Properties of Sulfur/Asphalt Binders

Viscosity. The viscosity of S/A binders and variations in this viscosity with temperature depend mainly on the sulfur content (Figure 2). The phenomenon of viscosity reduction is inverted as soon as the sulfur content exceeds 25–30% by weight.

The curves in Figure 3 correspond to modifications of the viscosity of binders according to temperature. These binders have a lower viscosity above 110°C than the corresponding asphalt under the same conditions. Similar tests were made on asphalts from various sources with penetrability figures of 180/220, 60/70, 40/50, and 20/30. The results obtained were similar to those quoted for the penetrability 80/100.

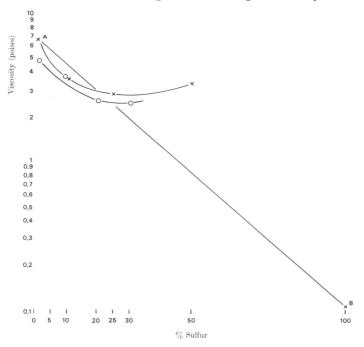

Figure 2. Viscosity of S/B emulsions with 80/100 bitumen at 130°C

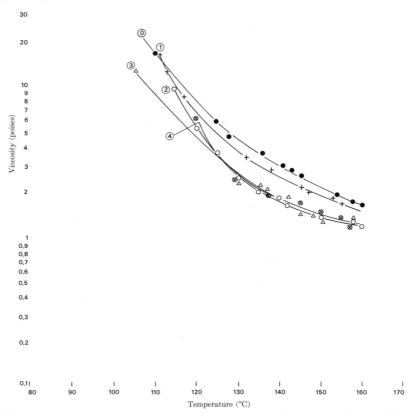

Figure 3. Variation of viscosity with temperature for S/B emulsions with 80/100 bitumen. ⊚ = pure bitumen, ● = 80/100. ① = 10% sulfur, + = 90% bitumen in wt. ② = 20% sulfur, ○ = 80% bitumen in wt. ③ = 30% sulfur, △ = 70% bitumen in wt. ④ = 40% sulfur, ⊗ = 60% bitumen in wt.

Specific Weight, Penetrability, and Softening Point. Specific weight at 18°C is measured by the Hubbard method. Penetrability at 25°C (100 g/5′) is determined according to ASTM method D 5 (NF 66004). The softening point is measured by ASTM method D 36.

Measurements were taken on binders prepared at 140°C, which were then stored at ambient temperatures for 15 days and raised under agitation to 160°C before the measurements were taken. This procedure was necessary, since the binders "matured" during the first 30 hrs, corresponding to reduced penetrability and increased softening temperature. It is suspected that this maturing is caused by a more important dissolution of dispersed sulfur. Binders prepared under these conditions, containing more than 30% sulfur by weight, are not sufficiently homogeneous for the measurements to be significant.

The following results were obtained (Figure 4).

1. The specific weight increased less than expected up to 10% sulfur contents, but increased normally thereafter.

2. Penetrability also increased up to 10% sulfur content. With over 20% sulfur, penetration reverts to its initial value, then decreases in almost linear fashion.

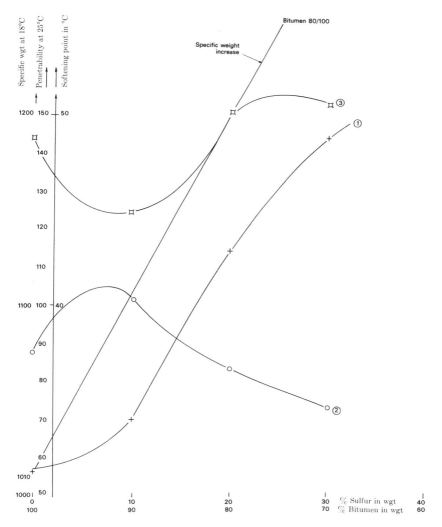

Figure 4. Specific weight, penetrability, and softening point of S/B emulsions. ① = *specific weight.* ② = *penetrability.* ③ = *softening point.*

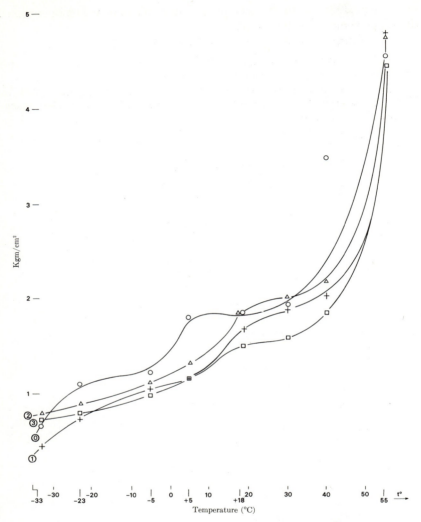

Figure 5. Cohesivity. ○ = *bitumen 60/70.* + = *10 sulfur.* △ = *20 sulfur.* □ = *30 sulfur.*

3. The softening point hits its minimum at a 10% sulfur content, exceeds the value for pure asphalt at 20%, and continues to increase thereafter.

4. It has been observed that maybe more than 14% sulfur by weight will dissolve in asphalt at 150°C. At 18 or 25°C the dissolved sulfur does not appear to have precipitated, since it would then perform in the same way as a loading agent—to reduce penetrability and increase the softening point.

5. The hypothesis of a sulfur/asphalt chemical reaction without release of hydrogen sulfide has been put forward by certain writers (3, 4) who have used vulcanization accelerating agents to reduce the "maturing" time to a few hours. After the chemical determinations of dissolved sulfur, this laboratory agrees that a sulfur/asphalt chemical reaction occurs.

6. It was observed that asphalt containing 10–20% sulfur has characteristics which differ from pure asphalt: specific weight: 1,040, penetrability: 102, softening point: 45. This medium serves as a matrix for the fine sulfur dispersion. Emulsions corresponding to increasing levels of sulfur content show an increase in their specific weight and softening

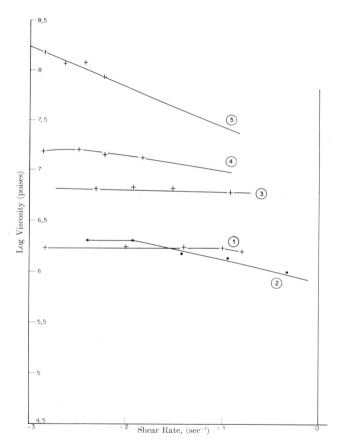

Figure 6. Variation of viscosity with shear rate (T = 25°C). ① = 80–100 conventional asphalt (linear). ② = 20–80 S/A emulsion with 80–100 asphalt. ③ = 40–50 asphalt (linear). ④ = 30–70 S/A emulsion with 80–100 asphalt. ⑤ = 20–30 blown asphalt (not linear).

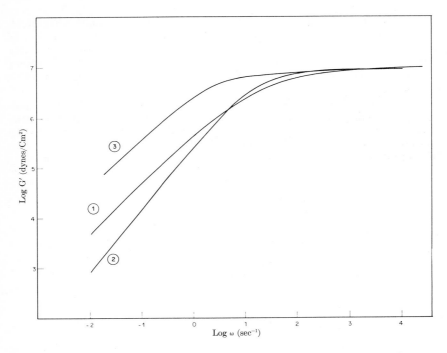

Figure 7. Elastic modulus G' (T = 25°C). ① = *80–100 shell asphalt.* ② = *20–80 S/A emulsion.* ③ = *30–70 S/A emulsion.*

point while their penetrability falls. Therefore entirely new binders with high sulfur contents are obtained.

Cohesivity. Cohesivity of the binders was measured using the Vialit method. The principle of this measurement process is described below.

A 1 mm layer of emulsion is used to join metal cubes on grooved faces having a surface area of 1 cm². After the assembly has been raised to the desired temperature, the lower cube is fixed on the fixed post of a pendulum head. A first measurement is made of the level to which the pendulum arm is raised, which gives the total energy absorbed. The cube pulled off during this measurement is then replaced on the base cube, and a second measurement is taken which corresponds to the total energy absorbed other than that required for rupture. The difference in the two levels to which the pendulum arm is raised is proportional to the energy absorbed for rupture of the emulsion layer at the temperature in question.

The rupture must occur in a plane which is parallel to and equidistant from the two faces to which the emulsion adheres, and the emulsion must not tear off the cube faces (hence the grooves).

The results of these measurements, carried out at between −33 and +50°C, are expressed in kg/cm². Sulfur/asphalt emulsions do not show marked changes in cohesivity, and measurements are generally dispersed. The cohesion curves are slightly below those for pure asphalt pen. 60/70 or 80/100 (Figure 5). For asphalt with penetration figures of 40/50 and 20/30, the measurements demonstrate no fresh elements, but the curves corresponding to the emulsions are slightly above that for the pure reference asphalt.

Viscoelastic Properties. Viscoelastic properties of S/A binders have been studied on Weissenberg rheogoniometer. Figure 6 shows the variation of viscosity coefficient with shear rate for some binders.

The 20/80 S/A binder (20% sulfur by weight) is not very different from the pen. 20/100 pure asphalt. But the 30/70 S/A binder is at the same time more viscous and not so linear (Newtonian) as the corresponding pure asphalt.

As a matter of fact, properties of 30/70 S/A binders with pen. 80/100 asphalt base lie between those of conventional pen. 40/50 asphalts and "blown" pen. 20/30 asphalts.

Figures 7 and 8 are the master curves at 25°C for elastic modulus G' and loss modulus G''. The relaxation distribution function has also

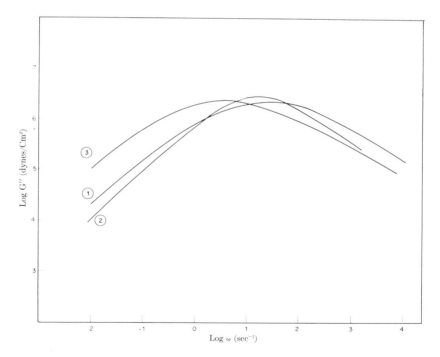

Figure 8. Loss modulus G'' ($T = 25°C$). ① $= 80$–100 shell asphalt. ② $= 20$–80 S/A emulsion. ③ $= 30$-70 S/A emulsion.

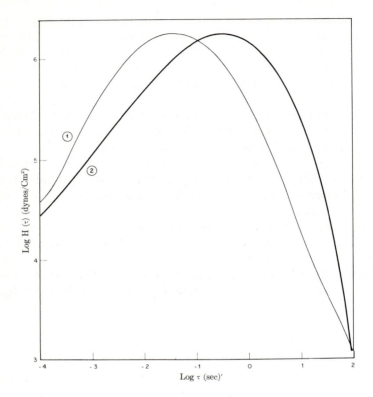

Figure 9. Relaxation distribution function using the Roesler and Twyman method. ① = 80–100 shell asphalt. ② = 30–70 S/A emulsion.

been calculated using the Roesler and Twyman method (Figure 9). The analysis of Figure 8 functions for this field of frequencies is shown in Table I. These results show that viscoelastic properties of 30/70 S/A binders with pen. 80/100 asphalt base are the same as those of pen. 40/50 "blown" asphalts (not represented graphically).

Table I. Loss Modulus G″

	Average Relaxation Time (sec)	Elasticity Coefficient (10^6 dynes/cm²)[a]	Viscosity Coefficient (10^6 poises)[a]
Pen. 80/100 asphalt	0,3	3,7	1,1
30/70 S/A binder	1,4	3,9	5,5

[a] Elasticity and viscosity coefficients are the sum of the coefficients of each element of the generalized Maxwell model. These results show that viscoelastic properties of 30/70 S/A binders with pen. 80/100 asphalt base are the same as those of pen. 40/50 "blown" asphalts (not represented graphically).

Sulfur/Asphalt Paving Mixtures

Sulfur/asphalt binders prepared under the conditions described above have special characteristics and are good binders for roads.

S/A Binders in High Performance Mixtures. Binders are used instead of pure asphalt to prepare familiar types of mixtures such as the semi-granular formula established by Setra and shown in Table II. A constant volume of binder was also used, but results are not shown graphically.

Table II. Setra Formula Mixtures

Pure Asphalt	S/A 10/90	S/A 20/80	S/A 30/70
0/2 Vignats 36	0/2 36	0/2 36	0/2 36
2/6 Vignats 32	2/6 32	2/6 32	2/6 32
6/10 Vignats 32	6/10 32	6/10 32	6/10 32
Asphalt pen.	S/A binder	S/A binder	S/A binder
80/100 6.6	10/90 6.6	20/60 6.6	30/70 6.6

The following measurements were made on Setra formula mixtures.

1. Compactibility.
2. Stability and performance under the influence of water using standardized methods: Duriez, Bresilien, and Marshall tests (*see* Appendix).

SPECIFIC WEIGHT. The specific weight of the test sample (Figure 10) increases according to the sulfur content of the binder during tests at constant weight. Compaction at 50, 75, or 100 blows/face has a secondary influence. The global curves have been represented graphically. The specific weight of the test samples varies little during the tests at constant binder volume, regardless of the sulfur content of the binder and the processing of the test sample. However, a greater dispersion of the results was noted but not represented graphically.

COMPACTIBILITY. The compactibility (Figure 10) of the test samples does not vary much with respect to the reference tests. Despite the measurement dispersion, compactibility decreases slightly when the sulfur content of the binder is increased for tests at constant binder weight. The tests at constant volume of binder are not shown graphically.

DURIEZ STABILITY FIGURES. For mixtures at constant binder weight these figures vary according to the sulfur content of the binder (Figure 11). Having touched a low minimum for a 10% sulfur content, the three curves rise sharply, reaching a value almost double the minimum (10% sulfur) at the 30% sulfur content level. There is a marked effect, although without any inversion of trend, of water on binders with 20 and 30% sulfur content, which furthermore follows the compactibility curve.

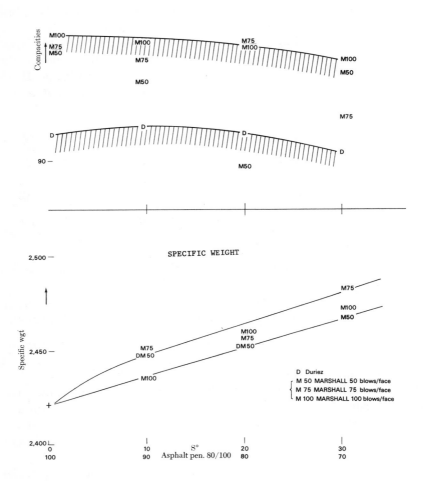

Figure 10. (top) Compactibility of mixtures, (bottom) specific weight

Bresilien Stability Figures. For mixtures with constant binder weight (6.6 parts) these figures vary according to sulfur content (Figure 12). A minimum level, still lower than that obtained with the Duriez tests, appears between 10–20% sulfur. Beyond this point, the curves rise sharply. The effect of water is marked, as with the Duriez tests, and follows the compactibility curve although there is no change in trend. As with the Duriez tests, this test shows a substantially improved result after 7 days, contrary to the sulfur-free product.

Marshall Tests. These tests (Figure 13) were carried out first in conventional form, at a rate of 50 blows/face for binders with increasing sulfur contents. The curve obtained was slightly downwards in trend. However, the test samples appeared inadequately compacted.

Tests were then carried out at 75 and 100/blows face. The improvement in compressive strength which then appeared for the pure asphalt binder was considerably less than that for products containing sulfur. With this type of granulometry it seems necessary to adjust the compacting conditions very carefully to obtain the best results.

FATIGUE TESTS. These tests are currently being made, and as yet there are no results to report.

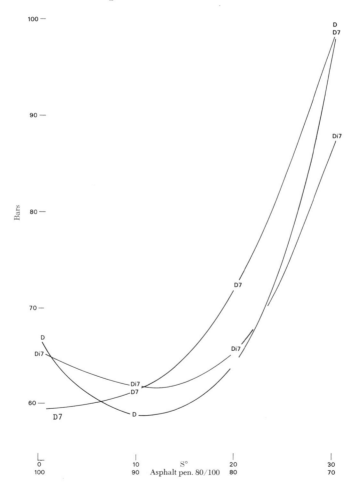

Figure 11. Duriez stability. D = immediate stability, D7 = stability after 7 days, Di7 = after 7 days with water.

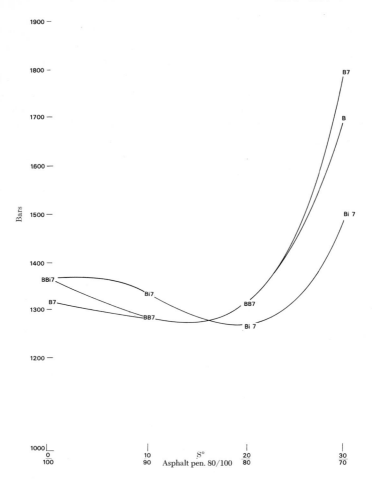

Figure 12. Bresilien stability. B = immediate stability, B7 = stability after 7 days, Bi7 = after 7 days with water.

The use of S/A binders to prepare high performance mixtures in accordance with Setra semi-granular formulation presents no particular difficulties and can provide products possessing roadway application properties which are superior to those of corresponding products having no sulfur content in their binders. The importance of the level of sulfur content has been shown as well as the highly marked influence of compacting in this case. Conventional methods have been used to test the mixes made with S/A binders. However, it is now recognized that it is necessary to adjust these conventional methods to get reproducible results.

S/A Binders in Low Quality Mixtures. The nature of the aggregates, their angularity, and their granulometric curves have a marked influence on the properties of mixtures in which they are used. Nevertheless, for reasons of availability and economics, mixtures are frequently prepared using aggregates with low inherent strength, which are round, show poor granulometric distribution, but are still adequate for the base layers.

The performance of mixtures prepared from S/A binders and low quality aggregates has been measured.

BASE LAYER SANDS (0/5). The 6 S/A formulation was studied (Tables III and IV) and shows high Marshall and Duriez compacity

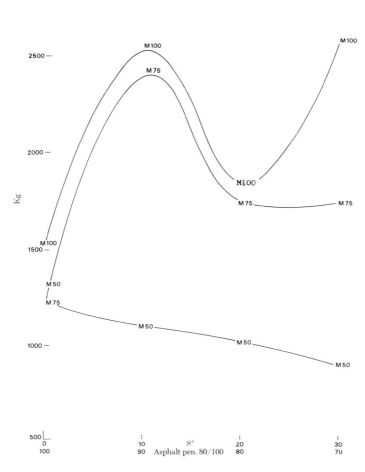

Figure 13. Marshall stability. M50 = 50 blows/face, M75 = 75 blows/face, M100 = 100 blows/face.

Table III. 6 S/A Formulation[a]

Landes sand	0/1	80 parts
Crushed material	0/4	20 parts
S/A	30/70	4.5 parts

[a] 30% of sulfur by weight. 70% of pen. 40/50 asphalt.

Table IV. Characteristics of the Base Layer[a]

	6 S/A	6A
Compacity (%)		
Duriez	75	71
Marshall	80	72
Stability		
Immediate	24	33
Duriez (bar)—7 days	52	31
Duriez (bar)—7 days with water	13	20
Immediate	10	7
Bresilien (bar)—7 days	14	12
Bresilien (bar)—7 days with water	4	6
Marshall—(Kg)	1040	

[a] 6A = pure pen. 40/50 asphalt.

figures with respect to the 6A pure asphalt reference. 6 S/A stability at 7 days exceeds that of the corresponding 6A grade. The Bresilien test results are substantially improved. A marked increase in performance with time and behavior at 40°C should be studied.

For asphalt sands, compressive strength at about 40°C is only one third of its value at 18°C and tensile strength falls to one tenth of its 18°C level. It must be shown whether products incorporating S/A binders are less sensitive or not to the effects of temperature variations.

BASE LAYER GRAVELS (0/25). The binders have a pen. 40/50 asphalt base, with the sulfur/asphalt ratio of 30/70 by weight. The formulations studied corresponded to a bituminous, semi-crushed type, concrete for base layer applications for a traffic load of 3 tons. The crushed material in these formulae was replaced by round material, and the filler was

Table V. Granulometries of Base Layer Gravels

Pure Asphalt 1A			2 S/A			3 S/A			Pure Asphalt 2A		
0/4	C	20	0/4	C	20						
0/5	R	20	0/5	R	20	0/5	R	30	0/5	R	30
6/18	C	30	6/18	C	30						
5/15	R	14	5/15	R	14	5/15	R	42	5/15	R	42
15/25	R	16	15/25	R	16	15/25	R	28	15/25	R	28
Chalk		2	Chalk		2						
Asphalt pen. 40/50		4	S/A 30/70		4	S/A 30/70		4.5	Asphalt pen. 40/50		4.5

Table VI. Test Results Using Formulations with Base Layer Gravels

		1 A	2 S/A	3 S/A	2 A
Compacity (%)					
Duriez		93	91	92	89
Marshall		94	92	94	92
Stability					
Duriez (bar)	immediate	82	66	66	53
Duriez (bar)	7 days		78	67	53
Duriez (bar)	7 days with water		46	43	41
Bresilien (bar)	immediate	30	15	42	20
Bresilien (bar)	7 days	29	17	43	17
Bresilien (bar)	7 days with water	21	11	30	12
Marshall (Kg)		1227	1260	1154	1100

eliminated when a S/A binder was used. Successive granulometries were obtained (Table V), and tests were carried out on these formulations (Table VI).

Tests made with pure asphalt (formulation 1A, semi-crushed) gave results slightly better than those for corresponding 2 S/A formulations. With the 3 S/A formulation, using only round material and no filler, the minimum performance level required for a surface layer was just reached, with allowance for the possible effect of ageing. The results, as a whole, are better than those for the 2 S/A formulation. The corresponding pure asphalt 2A formulation does not attain the minimum values.

SURFACE LAYERS (0/15). The binders used were pen. 40/50 pure asphalt and 30/70 and 20/80 S/A binders with pen. 40/50 asphalt base. Formulations were derived from the Setra semi-crushed range (Table VII), and were tested, giving results shown in Table VIII.

Product 7A with a pen. 40/50 pure asphalt base and a granulometry identical to that of 7 S/A and 9 S/A does not reach the average performance level of the S/A products. Product 6A with a pen. 40/50 asphalt base, although using crushed material only, produces results scarcely better than the corresponding semi-crushed S/A grades. The S/A semi-crushed formulations give results which meet the standard defined by Setra for surface layers. These test results justify the use of S/A binders

Table VII. Surface Layer Formulations Studied

9 S/A		7 S/A		7 A		6 A	
0/4 C	20	0/4 C	20	0/4 C	20	0/4 C	40
0/5 R	20	0/5 R	20	0/5 R	20		
6/12 C	30	6/12 C	30	6/12 C	30	6/12 C	60
5/15 R	30	5/15 R	30	5/15 R	30		
Chalk	1	Chalk	1	Chalk	1	Chalk	1
S/A 20/80	6	S/A 30/70	6	Asphalt pen. 40/50	6	Asphalt pen. 40/50	6

Table VIII. Test Results from Surface Layer Formulations

		9 S/A	7 S/A	7A	6A
Compacity (%)					
Duriez		96	93	90	95
Marshall		95	92	93	93
Stability					
Duriez (bar)	immediate	96	71	77	108
Duriez (bar)	7 days	106	83	75	104
Duriez (bar)	7 days with water	79	68	63	87
Bresilien (bar)	immediate	26	21	23	32
Bresilien (bar)	7 days	28	23	24	29
Bresilien (bar)	7 days with water	23	20	22	27
Marshall (Kg)		1050	1067	1172	1200

for surface layers associated with semi-crushed aggregates (less costly).

Formulations with a S/A binder base show a stability which always increases in the open air after 7 days. This is not the case with the corresponding pure asphalt base formulations. This improvement continues, following a regular pattern. Tests were not sufficiently numerous to define the period at the end of which, in all cases, measurement stability is obtained. But it is already known that the increase in stability value with time can be substantial.

Experimental Site

On-site tests were carried out on the perimeter road of the Lacq Plant 30 km west of Pau in the department of Pyrénées Atlantiques. The vehicle count on this road averaged 1,500 vehicles/day, of which about 50% were trucks.

Excavation work was first carried out beyond the mid-way line to widen the existing two-lane roadway (each lane was 4.2 m wide) for 1 km. The excavated half of the roadway was laid with foundations which were deliberately inferior to those of the unexcavated roadway. Base layers and surface layers were applied to the subbase layers with different formulations (Figure 14).

Preparation of the S/A Binders. The main difficulties arose from the poor thermal insulation of the pure sulfur circuits. Once these problems had been overcome, S/A binder was produced at a rate of 6 tons/hr. The emulsion had the correct sulfur content and satisfactory dispersion. Process parameters were: sulfur temperature, 140°C; pen. 40/50 asphalt temperature, 155–160°C.

Preparation of the Mixtures. It was possible to locate the storage tanks and all the materials necessary for the work on site around the mixing station. The station used is a Barber Green type 845 model, with an

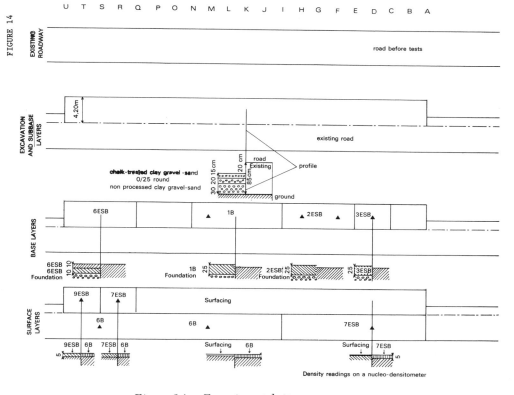

Figure 14. Experimental site

hourly output capacity of about 50 tons. This equipment presented no difficulties when adjusted for use with the S/A binder.

Comments on the Site. SAFETY. Hydrogen sulfide and sulfur dioxide measurements were taken at all points where the sulfur-laden products were handled using a Draeger type 21/31 manual pump unit and the MSA No. 2 explosivimeter. The levels of potential safety hazards are shown in Table IX.

The 50 ppm hydrogen sulfide peaks appearing on the finisher when the trucks tip their loads does not constitute a handicap. This problem

Table IX. Levels of Potential Safety Hazards at the Experimental Site

	H_2S (ppm)	SO_2 (ppm)	Explosivity (%)
Mixer	5	2	0
Trucks	5	2	0
Finisher	5 peaking to 50	2	0
Heated storage tank	100	2	60

is solved by providing additional fresh air in the driver's cabin. All on-site tests with the explosivimeter were negative.

On the heated storage tank, high hydrogen sulfide and explosivity levels were noted. This problem was satisfactorily solved by using an inert cover gas (nitrogen). A high explosivity level (close to 50%) was also observed on the storage tank with pure asphalt under the same conditions.

For these first road trials, it was difficult to obtain from workers the right conditions for preparing S/A binders and mixtures. Work at temperatures below 140°C is recommended to avoid gaseous effluents.

STABILITY OF S/A BINDERS. All test samples of the product taken from storage demonstrated the stability of the binders, *i.e.*, constant sulfur content. After cooling, reheating, and homogenization by pumping, no variation in sulfur content level was observed at various points on the circuit.

MIXING. S/A binders can be substituted without difficulty for pure asphalts and produce excellent mixtures. A marked reduction in absorbed energy was observed where S/A binders were used despite the fact that preparation was carried out at a lower temperature.

APPLICATION. In general, application temperature was less than 130°C when S/A binders were used. With the gravel–sand (2 S/A), application with the Barber Green finisher was carried out at 120°C in the first case, and rolling with a wet, smooth roller presented no problem. In the second case, application was at 145°C, followed first by pneumatic rolling (7 runs) and compacting with the smooth roller (10 runs). The optimum temperature for compaction (100°C) was reached 2½ hrs after application, but excellent compaction was still possible at 80°C.

With the sands (6 S/A) application at 120°C compaction was not possible. It was necessary to allow the temperature to drop to 100°C for shaping and to 80°C for elimination of all deformation using the smooth or pneumatic roller. During another test involving 6 S/A the smooth roller was run after waiting for 6 hrs. The temperature was then 70–80°C. The pneumatic roller was then applied. After 12 hrs grader operation was still possible.

With 3 S/A round gravel–sand only, the tendency was further accentuated. Compaction was impossible at over 100°C, but at this temperature it was necessary to commence work with a wet, smooth roller. A test was also carried out on a non-compacted surface applied with the finisher.

For the 7 S/A (Setra semi-crushed) surface layers, application was possible at 140°C, and compaction (two runs with the smooth roller, six runs with the pneumatic roller, and two further runs with the smooth roller) presented no difficulty. With formulation 9 S/A, application was at 120°C under conditions of heavy rain but without major difficulty. Compaction was still possible at 80–90°C (six runs with the pneumatic and four runs with the smooth).

These tests were carried out during June and July 1973, and the experimental site will be further observed in collaboration with the Bordeaux Laboratory of the Highway Administration.

First Conclusions

The uses of S/A binder mixes in pavements which are the same as for conventional mixes in overlays or new pavement construction, offer many economic advantages.

For a similar aggregate a significant portion of the asphalt (30%) can be replaced by sulfur. Total binder content of the mix, however, is similar to that normally used for asphalt alone. These results are obtained with conventional mix plant and paving equipment although with different directions for use.

Hot-mix paving mixtures are prepared at a lower temperature (30°C below the usual temperature) because of the lower viscosity of S/A binder. This results in lower energy consumption, saving 30% fuel and using less electric energy.

Application of hot-mix paving mixtures is performed more easily with conventional equipment. Moreover, the distance of truck transportation is improved, because of the lower compacting temperature, which can be below the melting point of sulfur. In addition, with some formulations, compaction is not necessary.

For similar pavement characteristics the use of this new S/A binder allows conventional aggregates to be replaced by low quality aggregates which are available on the spot. Moreover, with graded aggregates, the thicknesses of layers can be reduced as a result of the better characteristics of pavement. This is particularly interesting for the strengthening works.

The use of a thin layer of S/A binder mixture under the surface dressing allows solution of the maintenance problems of surface layers, *i.e.*, frequent damage of surface dressings subjected to hollowing and sweating phenomena is avoided, and levelling and rutting problems can be solved without having to wait for the placement of a conventional bituminous concrete.

The evaluation of the characteristics of the test sections (structural behavior, fatigue, etc.) is being conducted by the French Highway Laboratories. Definite conclusions will be made after they publish their results.

Appendix

Duriez Test (7). Appreciation of the mechanical property and resistance to mixture-breakdown under the effect of water, on cylindrical, compacted test samples of the mixture. The test is carried out at 18°C.

PREPARATION OF THE TEST SAMPLES. For granulometry less than 14 mm, the test sample weight = 1,000 g and the diameter = 80 mm. Double-acting compaction is at 120 bar in cylindrical molds for 5 min with a hot mixture (160°C). For granulometry greater than 14 mm, the expanded Duriez test is used. The test sample weight = 1,000 g and the test sample diameter = 120 mm. Double-acting compaction is at 120 bar in cylindrical molds for 5 min with a hot mixture (160°C).

MEASUREMENTS.

Single (axial) compressive strength. Speed = 1 mm/sec. Test immediately after 7 days dry and after 7 days of immersion.

Diametric compressive strength (Bresilien test). Speed = 1 mm/sec. Test immediately after 7 days dry and after 7 days of immersion.

Compactibility. Determine the apparent hydrostatic density of the test samples.

Water absorption. Determine the volumetric expansion.

Marshall Test (7). Measurements on test samples compacted to a given level, of the stability and flow-characteristics of hot mixtures not incorporating aggregates with a granulometry in excess of 20 mm.

PREPARATION OF THE TEST SAMPLES. Hot mixture is placed in cylindrical molds, having a diameter of 101.6 mm and is compacted by a 4.536 kg hammer descending in free fall from a height of 457 mm onto a metal base which has the same diameter as the mold and is placed on top of the mixture. The Marshall compacting process corresponds to 50 blows on each face. It can be complemented by compacting at 25, 75, and 100 blows/face. The mold must be placed on a wooden pedestal, and the 50 blows should be applied within about 55 sec.

MEASUREMENTS.

Stability. Compression test, following the generator of a semi-hooped test sample, after 30 minutes immersion in water at 60°C. Application speed = 0.86 mm/sec. Bend radius of the machines = 50.8 mm.

Flow-characteristic. Deflection of the test sample up to the point of rupture.

Literature Cited

1. Bacon, Bencowitz, *Book ASTM Stand.* (1938) **38** (2), 539.
2. Bacon, R. F., Bencowitz, I., U.S. Patent No. **2,182,837** (1936).
3. Quarles *et al.*, *Brennsto. Chemie* (1962) **43** (6), 173.
4. *Ibid.*, (1965) **45** (1), 1.

5. Petrossi et al., *Ind. Eng. Chem. Prod. Res. Develop.* (1972) **11** (2), 214.
6. Bocca et al., *Chim. l'Ind. Milan* (1973) **55** (5), 425.
7. "Nouveau Traite de Materiaux de Construction," Duriez et Arrambide, Dunod Edit, 1961.

RECEIVED May 1, 1974

9

Civil Engineering Applications of Sulfur-Based Materials

B. R. GAMBLE, J. E. GILLOTT, I. J. JORDAAN, R. E. LOOV, and M. A. WARD

Department of Civil Engineering, University of Calgary, Alberta, Canada T2N 1N4

> *The current surplus of elemental sulfur, principally in Western Canada, raises the possibility of using sulfur or materials using sulfur as a binder in a variety of civil engineering applications. The strength of sulfur varies indirectly with the hydrogen sulfide concentration. The creep rate of sulfur at room temperature is high and is temperature dependent. As expected, the creep rates of composites containing sulfur as a binder are also relatively high compared with conventional concretes. Depending on the mineralogical composition of the soil, and in the absence of swelling clays, sulfur has potential for soil stabilization. Lastly, a lightweight sulfur material was produced using fly ash. Other additives were used to create foams with limited success.*

Since 1968, sulfur production, principally as an involuntary by-product of sour gas processing in Western Canada and other parts of the world, has exceeded demand. Such market pressures have depressed the price of sulfur from a high of $35 to a current price of $13/ton at producing plants in Alberta. Figure 1 illustrates the current and projected supply and demand situation.

Given an abundant supply at a low price, sulfur may find applications in civil engineering which take advantage of one or more of its interesting chemical and physical properties (1). For example, sulfur may be used in insulation because of its particularly low coefficient of thermal conductivity. Alternatively, because of its high compressive and bond strengths and the ease with which aggregates or fillers can be mixed with sulfur in the molten state, the element is of interest in structural situations. This may be of special importance where the resistance of

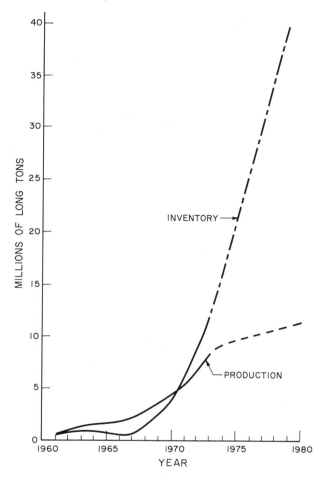

Figure 1. Canadian sulfur production and inventory

sulfur to those salts and acids which are deleterious to portland cement concretes can be exploited. When sulfur is used in an asphalt emulsion, the physical properties of the binder are improved.

Unfortunately, some of the properties of sulfur can also cause problems. In many potential applications, especially those in which sulfur is to be used as a binder, the need to heat the entire mix to temperatures of 140°C and above is a serious drawback. Also, when casting a material containing molten sulfur, freezing of the melt is accompanied by a significant amount of shrinkage which requires special consideration. A further problem is illustrated by Figure 2; the combination of a low thermal conductivity and a high coefficient of expansion will lead to internal stresses when sulfur is subjected to rapid ambient temperature

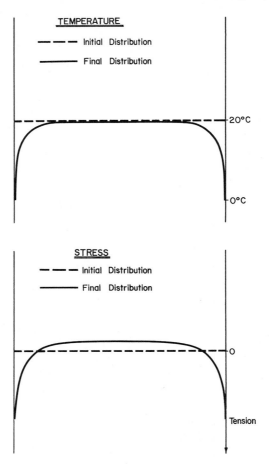

Figure 2. Temperature and stress distribution in a quenched sulfur specimen

changes. These problems are among those which will restrict the use of sulfur as a replacement binder for portland cement paste in conventional concretes.

Another difficulty facing the potential sulfur user is the dearth of engineering information regarding the properties of either sulfur itself, composites in which sulfur acts as a binder, or sulfur-containing composites. In each case there is a need to determine strength, elastic and time dependent strain properties, coefficients of thermal expansion and conductivity, indices of durability (alternate wetting and drying, thermal cycling, and performance in other deleterious environments), and the effects of impurities such as hydrogen sulfide or other deliberately added compounds. The suitability and advantages of various fillers and aggregates need to be established, together with methods of proportioning sulfur composites from the viewpoint of economy and property optimization.

It is clear that a wide variety of problems challenge researchers entering this interesting field. For the past 13 months the authors have been examining several of these problems, and this paper outlines the progress made to date. Since all aspects of the project are still active, the presentation will be necessarily incomplete until final results can be published at a later date.

For want of a well established terminology, the phrase "sulfur-bound composites" is used here to differentiate those materials in which sulfur is the main binder from "sulfur-containing composites" in which other materials, such as portland cement or asphalt, may act as the predominant binder.

Elemental Sulfur

In most potential applications sulfur will be combined with other materials. However, several factors will influence the properties of the composite *via* their effect on sulfur itself. In the case of sulfur-bound composites, factors affecting the strength and deformation characteristics of elemental sulfur are of particular interest. As an example, the rapid increase in strength with age after casting of sulfur or its composites has been observed in several investigations. The gain is believed to be mainly associated with the crystallographic inversion from the monoclinic to the more stable orthorhombic polymorph, although recrystallization may also be involved.

Effect of Hydrogen Sulfide on Strength. The introduction of just a small quantity of one of a number of impurities or additives may critically affect strength. Rennie *et al.*, have stated that hydrogen sulfide is one of these compounds (*2*). The tests described below were conducted to establish the significance of this effect.

Table I. Hydrogen Sulfide Levels[a]

	H_2S Level from Sample	
Condenser	Before Casting	After Casting (3-day specimens)
1	78.3 (5)[b]	44.1 (12)
2	77 (5)	44.2 (12)
3	11 (6)	7.0 (15)
4	2 (13)	1.0 (15)

[a] In ppm.
[b] Figures in brackets indicate the number of days between 3-day strength test and H_2S determination.

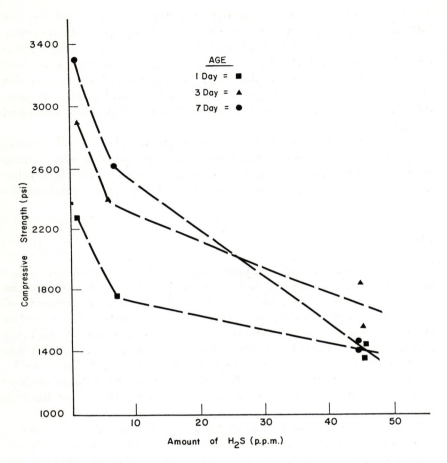

Figure 3. Hydrogen sulfide concentration vs. compressive strength

Four samples of sulfur from four different condensers were obtained from the Petrogas plant in Balzac, Alberta. At the plant the hydrogen sulfide level varied from 168 to 18.5 ppm. Table I indicates that substantially lower values were obtained in the laboratory before the samples had been remelted. A further reduction occurred as a result of remelting and storage.

Compressive strength was determined from 3 in. × 6 in. cylinders. The cylinders were capped immediately after casting to prevent cracking of the specimens. A sulfur-based capping compound normally used for concrete cylinders was used. Flexural strength was determined from 1½ in. × 1½ in. × 10 in. specimens using third point loading. It was found that cracking before testing could be avoided by casting vertically. At each age of 1, 3, and 7 days, three compressive specimens at each of four hydrogen sulfide concentrations were tested. Likewise, two flexural specimens were tested at 1 day and three at 3 and 7 days.

Results of compression and flexure tests are given in Figures 3 and 4, respectively. A statistical analysis confirmed that the effect of hydrogen sulfide concentration is significant. The magnitude of the strength variation is clearly of practical importance. For example, strength reductions up to 50% resulted from high hydrogen sulfide concentrations. It has yet to be determined whether fillers or additives will reduce or eliminate this effect or if other physical properties, such as elastic moduli or creep, will also be affected by the hydrogen sulfide content. Since hydrogen sulfide concentration appears to be such an important parameter, it should be determined and reported in any work involving the properties of sulfur.

Effect of Temperature on Strength. Work by Rennie *et al.* indicates that the flexural strength of 1½ in. × 1½ in. × 10 in. sulfur prisms depends markedly on the temperature at which they are equilibrated before testing (*2*). Other investigators (*3*) have noted that sulfur specimens having a substantial volume/surface area ratio may undergo "thermal shock," a durability problem in which a fluctuating temperature may

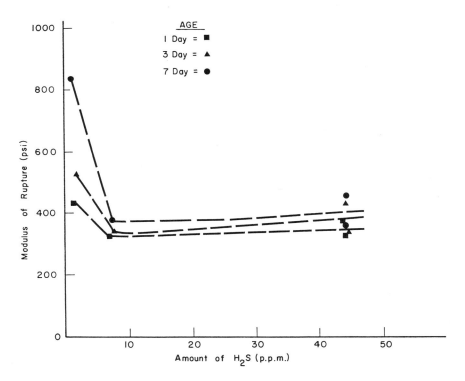

Figure 4. Hydrogen sulfide concentration vs. modulus of rupture

cause disruption with time. These would be important phenomena in many applications.

Researchers working to understand the causes recognize that a specimen size effect may exist. This would affect the time strength–temperature relationship and may also explain the durability problem. As mentioned in the introduction, internal stresses are developed in specimens subjected to ambient temperature changes. At a given conditioning temperature the magnitude of the effect depends on the size of the specimen, and likewise the measured strength also reflects this parameter. For a given specimen size, the dependence of the stress state on conditioning temperature may well mask the local strength–temperature relationship, which can only be observed in small specimens.

Creep and the Effects of Temperature. When sulfur or sulfur composites are subjected to sustained stress, creep is observed. Since the magnitude of this phenomenon is significant from the engineering point of view, the causes of creep are of interest.

The possible mechanisms are numerous and include movements caused by the transformation from one polymorphic form to another while under stress, disruptive or non-disruptive movements within crystals, between crystals, or between sulfur and the aggregates or fillers. Considerable work has yet to be done before the roles of these mechanisms are clarified. Thus far the experiments substantiate that creep is temperature dependent.

Sulfur-Bound Composites

Sulfur-cretes. In previous investigations by the authors the complete replacement of the portland cement paste component of normal concrete by sulfur or sulfur–fly ash mixtures was found to be feasible. Such results have been duplicated in the present study, and additional tests have shown that by using conventional concrete aggregates, potentially useful materials having sulfur contents as low as 10% by weight can be fabricated. Since these materials, known as sulfur-cretes, may find structural uses in situations where portland cement concrete is unsuitable, further study of strength, deformation, and durability behavior is in progress. Though this work is far from complete, preliminary results from sustained compressive loading tests show that a significant amount of creep occurs at room temperature (Figure 5).

Soils and Other Fillers. Sulfur may be of value if it can be used to give insulating and/or structural qualities to soils and other fillers which are readily available, but not usually considered suitable for construction purposes. A number of tests have been performed to assess the potential of sulfur in relation to such materials.

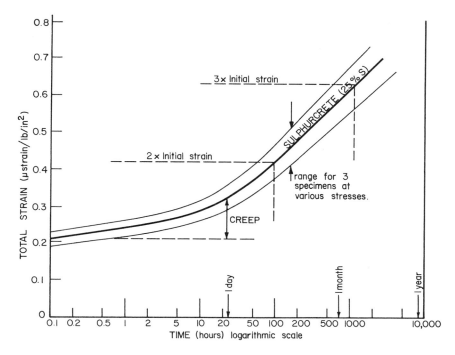

Figure 5. Creep behavior of sulfur at 70°F

Fillers which were combined with sulfur included the principal clay minerals as well as diatomaceous earth, fly ash, and gypsum. These materials were generally combined singly with sulfur to facilitate the interpretation of results, but in some instances more than one was used. Trials were made to establish the filler proportions which could be mixed readily with the molten sulfur (130°C), and these designs were used to fabricate test specimens (Table II).

Samples were fabricated in batches of at least 12, and values of both compressive and flexural strength were obtained on three specimens or

Table II. Mix Designs in Proportions by Weight

	Additive	Sulfur
Bentonite	27	73
Illitic shale	34	66
Kaolin	23	77
Fly ash	40	60
Diatomaceous earth	17	83
Mortar		
Illite	8	39
Ottawa sand	53	39

Table III. Compressive Strength of Sulfur–Additive Combinations[a]

Additive	Age (days)		
	1	3	7
Bentonite	5330 ± 166	5400 ± 81	5775 ± 118
Illitic shale	3850 ± 188	4000 ± 188	4300 ± 31
Kaolin	—	—	1500 ± 51
Fly ash	900 ± 44	1200 ± 62	1700 ± 62
Diatomaceous earth	1200 ± 60	1500 ± 38	1700 ± 38
Mortar	—	—	5700

[a] Results in psi expressed at 95% confidence level on mean.

Table IV. Rupture Modulus of Sulfur–Additive Combinations[a]

Additive	Age (days)		
	1	3	7
Bentonite	270 ± 31	320 ± 44	335 ± 53
Illitic shale	390 ± 38	400 ± 150	415 ± 44
Fly ash	350 ± 50	550 ± 50	580 ± 48
Mortar	—	—	900

[a] Results in psi expressed at 95% confidence level on mean.

more at each age (Tables III and IV). Samples were tested after periods of 1, 3, and 7 days from casting. Compression strength tests were made on cylinders measuring 3 in. × 6 in., and flexural strength tests were made by third point loading of 1 in. × 1 in. × 11 in. prisms. A few values of elasticity modulus were obtained. Results of the compression tests reported are for samples capped immediately after demoulding.

Very fine-grained materials combined readily in the proportions of about 1/3 filler to 2/3 sulfur, whereas these proportions were approximately reversed when sulfur was combined with both sand and clay (Table II). After the melt had cooled to room temperature, the strength rapidly increased during the first few days with little further change after about 1 week. Compressive and flexural strengths were of the same order as those of portland cement concretes (Tables III and IV). Among the fine-grained additives, the highest strengths were recorded on sulfur mixes made with bentonite and illitic shale. The low strength of the kaolin-containing samples may be attributed to the cracking which usually occurred on cooling.

A preliminary test of durability was made by immersing cubes of the various mixes in water. Mixes made with as little as 3% bentonite together with 30% illitic shale cracked after immersion for 1 day while samples containing 20% bentonite and 14% illitic shale disintegrated. The other sulfur-additive mixtures showed no sign of distress after immersion for several weeks (Figure 6).

In conclusion, results suggest that, depending on the mineralogical composition of the soil, sulfur may have a potential in soil stabilization for the subgrade of roads or airfields. It is evident, however, that the presence of a few percent of a swelling clay would make sulfur unusable in this application unless the clay mineral is made hydrophobic or moisture is excluded. When minerals of this sort are absent, the rapidity with which soil sulfur mixes gain strength may be advantageous. In cold regions, for example, where the low temperatures reduce the rate of chemical reactions responsible for strength gain in soil–cement and soil–lime combinations, sulfur may be a more suitable bonding agent. The simplicity of the mixing and forming operations also suggests that these mixes may find a use as a building material in certain circumstances.

Production of Sulfur Foams. The insulating value of sulfur composites may be enhanced and their density decreased by the production

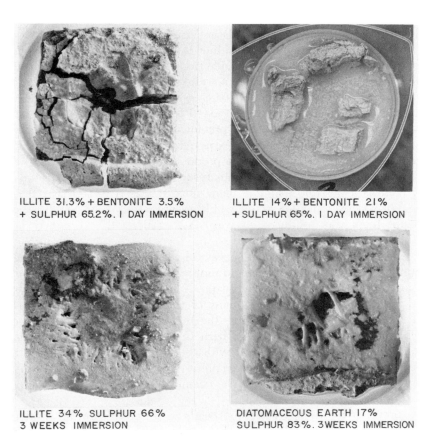

Figure 6. *Effects of water on sulfur with mineral additives*

of foam. This subject has been treated by Dale and Ludwig in a series of papers (4, 5) and by Woo and Campbell of Chevron Research Ltd. (6). During this investigation it was found that many naturally occurring substances which could be considered as filler materials will also cause foaming to some degree. The following is a partial list of materials which were used in this study: fly ash; Athabasca tar sand, coke, and tailings; gypsum; peat moss; and sheep manure.

When equal proportions by weight of fly ash and sulfur were mixed, a strong porous solid was formed. The resulting material weighed 95 lb/ft^3 and had a compressive strength of 1760 psi at 7 days.

Tar sand, pulverized coke, and tar sand tailings all caused foaming when mixed in molten sulfur. Unfortunately, it was not possible to stabilize these foams; in all cases the bubbles burst on heating and collapsed on cooling. In addition, the tar sand and coke segregated by floating to the top while the tailings segregated by settling to the bottom of the mould.

As has been reported by Dale and Ludwig (5), a lightweight foam may be produced by combining sulfur, talc, pinene, tri-cresyl phosphate, phosphorus pentasulfide and phosphoric acid. An attempt was made to produce a foam by using Calgon, a well-known dispersing agent containing phosphate. The pinene and tri-cresyl phosphate used by Dale and Ludwig was omitted from these mixes. It was found that the resulting foam was comparable with that previously produced. Two benefits of some possible interest were a sharp reduction in odor and a marked reduction in the cost of the foam-producing ingredients.

In contrast to the serendipitous results obtained above, gypsum was selected as a possible foaming agent since it was reasoned that the release of its crystallization water would create a foam. Although a lightweight foam was produced, it was very brittle with excessively large bubbles.

Based on information that successful glass foams had been produced through the addition of manure to molten glass, sheep manure was used in an attempt to create a sulfur foam. A substantial amount of foaming did indeed occur but, as with many of the other foaming agents tried, no means was found to stabilize the mix. Peat moss, a more readily available organic material in northern Canada, was also tried and gave similar results to those obtained with sheep manure. It is possible that the foaming on the addition of manure or peat moss resulted from the formation of steam rather than a chemical reaction with the organic material.

Smith *et al.* used polystyrene beads in sulfur to form an insulating material (7). Unfortunately, the beads are melted by the molten sulfur. It was believed that exfoliated vermiculite might serve the same purpose but difficulties were experienced because of the tendency of vermiculite to float to the surface of the mix.

Sulfur-Containing Composites

The insulating value of sulfur may justify its use as a partial or full replacement of other binders such as portland cement or asphalt in situations where the lowest possible thermal conductivity coefficient is desired. Such materials are potentially useful where pavements are designed with the "thermal blanket" concept in mind such as in permafrost areas. Alternatively, it may be necessary to prevent frost heave of the subgrade, and the thermal blanket may reduce heat losses during periods of cold ambient temperature. Dow Chemical (8) has attempted to solve these problems using foamed polystyrene insulation between the base course and the sub-base. Pronk (9) has obtained encouraging results on a 1500-ft test section in Calgary with a cellular lightweight aggregate as a replacement for crushed rock, using a conventional asphalt.

The incorporation of sulfur in asphalt emulsions also shows great potential. Although further improvements in the insulating value of asphalt might be expected by adding sulfur, such binders have other advantages over normal asphalt. In 1972 Pronk (9) produced sulfur asphalt emulsion mixes in the laboratory. More recently, Société Nationale des Pétroles d'Aquitaine (SNPA) (10) reported the results of an extensive laboratory and field study using sulfur in an asphalt emulsion. Both Pronk and SNPA have shown that temperature–viscosity relationships are improved, as are Marshall stability and flow values.

The extension of the above work to include mixes containing lightweight aggregate is the objective of the current study being conducted by Pronk and researchers at the University of Calgary, supported by the Sulphur Development Institute of Canada (SUDIC). Several mix designs have been developed for base courses using a wide range of sulfur–asphalt combinations. The thermal conductivity and durability of these mixes are being studied to determine optimum mix proportions for a subsequent field study. A test section will be laid using conventional paving equipment in late summer 1974. The test section will be located in an area with soil susceptible to frost heave. The pavement will be instrumented to monitor the thermal gradient through the pavement and subgrade.

Although present work is concentrating on thermal blanket applications, a concurrent study is concentrating on mixes containing crushed rock aggregates. If the price of asphalt continues to rise it is clear that in some areas it may become both feasible and desirable to use sulfur–asphalt emulsions in such applications.

Acknowledgments

The authors wish to acknowledge the considerable assistance of J. B. Hyne and co-workers of Alberta Sulphur Research Limited, particularly for the hydrogen sulfide analyses. R. Begert capably carried out many of the experiments under the careful supervision of Howard Johnson.

Literature Cited

1. Loov, R. E., Vroom, A. H., Ward, M. A., "Sulphur Concrete—A New Construction Material," *J. Prestress. Concrete Inst.* (Jan.-Feb. 1974) **19**, (1), 86–95.
2. Rennie, W. J., Andreassen, B., Dunay, D., Hyne, J. B., "The Effects of Temperature and Added Hydrogen Sulphide on the Strength of Elemental Sulphur," *Alberta Sulphur Res. Quart. Bull.* (Oct.-Dec. 1970) **7**, (3), 47–60.
3. Hyne, J. B., personal communication, 1973.
4. Dale, J. M., Ludwig, A. C., "Preparation of Low Density Sulfur Foam," Cold Regions Research and Engineering Laboratory Report **206**, Sept. 1967.
5. Dale, J. M., Ludwig, A. C., "Investigation of Lightweight Sulfur Foam for Use in Field Applications," Cold Regions Research and Engineering Laboratory Report **227**, Oct. 1969.
6. Woo, G. L., Campbell, R. W., "Sulfur Foam—A New Rigid Insulation," *Joint Chem. Eng. Conf.*, **4th**, Vancouver, British Columbia, Canada, September 9-12, 1973.
7. Smith, N., Pazsint, D., Karalius, J., "Laboratory Development and Field Testing of a Sulfur/Foamed Polystyrene Insulation Composite," *Joint Chem. Eng. Conf.*, **4th**, Vancouver, British Columbia, Canada, September 9-12, 1973.
8. "Insulated Highways for Canada's North," *Science Dimension* (1973) **5**, (4), 16–20.
9. Pronk, F. E., Reports to the City of Calgary Engineering Dept., Feb. 4 and Apr. 25, 1972.
10. Société Nationale des Pétroles d'Aquitaine, "Properties of Sulphur Bitumen Binders," International Road Federation World Meeting, **7th**, pp. 39, Munich, Oct. 1973.

RECEIVED May 1, 1974. This project was supported from a National Research Council of Canada Special Project Grant and a research contract held with the Sulfur Development Institute of Canada (SUDIC).

10

Cold Region Testing of Sulfur Foam and Coatings

JOHN M. DALE and ALLEN C. LUDWIG

Southwest Research Institute, San Antonio, Tex. 78284

> *Sulfur foams and coating compositions are potentially useful in cold regions. Several applications involve soil–sulfur contact where there is the possibility of sulfur degradation by soil microorganisms and other environmental effects. In a test program started in 1970, samples of foamed sulfur and sulfur coatings were buried in the soil at one location in Colorado and three locations in Canada. These samples were retrieved at yearly intervals and their physical properties compared with control samples kept in the laboratory. After 3 yrs of burial, no deterioration in strength could be measured. There was no evidence of bacteriological or chemical deterioration or of damage by thermal shock or freeze–thaw.*

During the past 12 yrs considerable attention has been given to developing rigid sulfur foams and protective coatings made largely from sulfur. In many of the potential uses, the materials are in direct contact with the soil. Sulfur foam may someday be widely used as roadway or runway subbase insulation either to protect the road subbase from freezing or to protect a permafrost subbase from thawing. In either case, the foam will be buried approximately 1 ft below the surface, deep enough not to be affected by the daily temperature cycling on the surface. Sulfur foam might also be used as subbase insulation for homes or cold storage warehouses. Sulfur coatings could be used to seal underground structures such as basements, where they are applied to exterior surfaces with a soil backfill or to line pits and tanks in which soil-like sediments accumulate. There are other potential applications where the materials will be both buried and exposed.

Outdoor weathering studies (1) indicated that there is negligible surface loss from sulfur films by oxidation under ambient natural atmospheric conditions. Accelerated weathering of sulfur films in the atmosphere has indicated that ozone, ultraviolet light, and water vapor are much less important factors than temperature. Above 80–90°C (175–195°F) sublimation losses become important. However, these temperatures are seldom if every encountered around normal building structures.

A variety of sulfur oxidizing microorganisms in the soil can convert elemental sulfur and other reduced forms of sulfur to the sulfate. This has agricultural importance because plants take up most of their sulfur as the sulfate. However, the action of these microorganisms can be quite undesirable in other instances, such as causing sulfur compounds used for sealing underground pipes to deteriorate. Some of the better known sulfur-oxidizing microorganisms are *Thiobacillus thiooxidans, Thiobacillus thioparus,* and *Thiobacillus denitrificans.* These and other sulfur-oxidizing microorganisms are found in virtually all arable soils and can function over a wide range of conditions.

There is a substantial body of literature relating to this activity (2). In one study (3) specimens of sulfur jointing compounds (cements) containing 0.1–1 wt % of various bactericides were prepared at 130–140°C and buried 18 in. below the surface in Pittsburgh, Penn. and 6 in. below the surface at Newgulf, Tex. The Newgulf site was an old sulfur storage vat site which was extremely contaminated by sulfur-oxidizing bacteria and where the soil pH was 3. After exposure for 5 yrs the specimens buried in Pittsburgh showed practically no weight change and only a slight loss in strength, whereas the specimens buried at Newgulf had lost considerable weight and nearly 50% of their strength. Bactericides retarded or prevented the action of bacteria on the sulfur cements.

Man's expanded activities in the arctic regions and the colder regions of the temperate zone are increasing the potential applications for sulfur foam and coatings in these areas. A research program (4) conducted by Southwest Research Institute (SwRI) for the United States Army Cold Regions Research and Engineering Laboratory on the Arctic Circle demonstrated that rigid sulfur foams and sulfur coatings can be produced in cold weather at a remote site. Unfortunately, there was no information available as to how these materials might stand up to the environment with time, so The Sulphur Institute decided to sponsor a program at SwRI to determine if these materials could withstand the enviromnent of the cold regions.

Specimen Preparation

The rigid sulfur foam specimens were prepared from a typical SwRI foam formulation containing approximately 80% sulfur, 10% talc, and 10% foam-forming additives and had an average density of 15 lb/cu ft.

Figure 1. Sulfur foam specimen

The individual specimens shown in Figure 1, measured 6 × 2 in. They were cut from board castings so that the internal cell structure was exposed on four sides. The coating specimens were prepared by joining two 1 × 2 in. concrete blocks end to end and with the molten coating applied to both the top and the bottom (Figure 2). The coating was a sulfur formulation containing 3% dicyclopentadiene and 1.5% glass fibers 0.5 in. long and was brushed on at 155°C. Half of all foam and coating specimens incorporated 0.5% sodium pentachlorophenate in the sulfur formulation. Sodium pentachlorophenate has been shown (5) to be a very good bactericide for controlling sulfur-oxidizing bacteria.

Test Areas

Twelve foam specimens (six with bactericide and six without) and 12 coating specimens (six with bactericide and six without) were placed at each location using the same burial method. A trench 1.5 ft wide × 1 ft deep × 6 ft long was dug, and the foam specimens were placed on edge and the block specimens laid flat. All specimens containing bactericide were placed so that they proceeded from the center of the trench in one direction, and those specimens without bactericide proceeded from the center of the trench in the opposite direction. The recovery sequence

Figure 2. Sulfur-coated block specimen

was designed to proceed from each end toward the middle. The last specimens were placed during the fourth week of September 1970. The first test site was provided by the Colorado Department of Highways and is adjacent to their Foam Roadway Subbase Test Site near the summit of Vail Pass, some 100 miles west of Denver. The pH of the soil at this site was 7.3. The second test site was provided by Texasgulf, near the entrance to their Okotoks Sulphur Recovery Plant, 23 miles south of Calgary, Alberta, Canada, in a plains area used for grazing. The pH of the soil at this site was 8.0. The third site was provided by the Alberta Department of Highways at their Canmore District which is 60 miles west of Calgary in the Canadian Rockies. The pH of the soil at this site was 8.5. The fourth site was provided by Texasgulf and was located on the grounds of their plant site at Whitecourt, Alberta, Canada, approximately 100 miles northwest of Edmonton. The physical location of the burial site at this location was in the sulfur storage vat area where there was evidence of bacterial activity and, except for its northern location, it had many of the characteristics of the test site used in the previous study at Newgulf, Tex. The pH of the soil at this site was 6.6. A set of control specimens was kept at SwRI.

Appraisal of Recovered Specimens

The test specimens were placed during the summer of 1970 and recovered during the summers of 1971, 1972, and 1973. Test specimens at all four locations were removed only in the summer because of frozen ground conditions in the winter.

The necessary process of packing the specimens, transporting them to the test sites, placing them in the excavation, filling in, compacting the soil, digging them up, and transporting them back to the laboratory is abrasive in itself. The added inspections by customs and more recently by airline security personnel increased the probability of damage to the specimens. Fortunately, the test specimens were large enough that it was possible to cut smaller sections from the damaged specimens sufficient to conduct all the physical tests on all the foam specimens and on all but three of the coating specimens. The specimens were returned to the laboratory, cleaned, and allowed to equilibrate. Three test specimens were cut from each of the foam specimens, the compressive strength of each was measured, and the average was recorded. These data are shown in Table I.

The strength measurements of the third-year foam specimens from the Vail Pass, Colo. test site indicated lower compressive strengths. In checking the data on these specimens it was found that they had a lower density when made. The measured values shown in Table I are typical

Table I. Compressive Strength of Sulfur Foam Specimens

Test Location	Foam Specimen No.	Without Bactericide (psi)			With Bactericide (psi)		
		1st yr	2nd yr	3rd yr	1st yr	2nd yr	3rd yr
San Antonio (Control)	SA	52	51	52			
San Antonio (Control)	SA-B				31	40	55
Whitecourt, Alberta, Canada	CZ	45	35	40			
Whitecourt, Alberta, Canada	CZ-B				29	55	54
Okotoks, Alberta, Canada	CX	47	51	50			
Okotoks, Alberta, Canada	CX-B				33	31	29
Canmore, Alberta, Canada	CY	45	44	41			
Canmore, Alberta, Canada	CY-B				41	38	36
Vail Pass, Colo.	D	49	49	37			
Vail Pass, Colo.	D-B				49	57	26

for materials of that density and are thus not the result of any environmentally induced damage. The coating specimens were tested by measuring their flexural strength both at the joint and through the block. These data are shown in Table II.

When specimens were removed at the end of the first year, we noted at both the Okotoks and Vail Pass sites that roots of surrounding vegetation had grown up to the surface of the specimens, and it appeared that they might grow into them, particularly the foam specimens where the cell structure was exposed. When specimens were removed at the end of the second and third year, we found that roots had not grown into the specimens, but had proceeded laterally along the surface. This is in contrast to some organic foam materials which readily host root growth.

When removed, all specimens were closely inspected for signs of bacterial or chemical degradation from possible reactions with soil constituents. Nothing of this nature was observed, and after light cleaning the specimens had, for all practical purposes, the same appearance as when they were buried. The specimens removed were also inspected for thermal shock and freeze–thaw degradation. Thermal shock failure in sulfur materials is shown by large cracks through the specimens and is caused by large and sudden temperature variations while freeze–thaw failure in sulfur is evidenced by many small cracks and surface spalling. Neither of these failures was observed.

Table II. Flexural Strength of At Joint

Test Location	Foam Specimen No.	Without Bactericide (psi)			With Bactericide (psi)		
		1st yr	2nd yr	3rd yr	1st yr	2nd yr	3rd yr
San Antonio (Control)	SA	480	355	615			
San Antonio (Control)	SA-B				370	442	525
Whitecourt, Alberta, Canada	CZ	815	795	535			
Whitecourt, Alberta, Canada	CZ-B				410	600	490
Okotoks, Alberta, Canada	CX	640	560	685			
Okotoks, Alberta, Canada	CX-B				455	—	—
Canmore, Alberta, Canada	CY	435	615	600			
Canmore, Alberta, Canada	CY-B				460	355	—
Vail Pass, Colo.	D						
Vail Pass, Colo.	D-B						

The lack of any evidence of bacterial activity at any of the test sites may be related to several factors. The bacterial oxidation of sulfur reportedly can occur from about 4 to 55°C (39–131°F) with the most favorable temperatures being between 27 and 40°C (80–104°F). Thus, the temperature at the test locations was not conducive to bacterial activity. Bacterial oxidation of sulfur has been shown (2) to increase with decreasing sulfur particle size and is enhanced by mixing the sulfur with soil to improve the soil–sulfur contact. However, sulfur foams and sulfur coatings are massive forms of sulfur and do not have good soil–sulfur contact. Finally, the additives used in preparing the foams and the coatings may have had some bactericidal influence which could not be identified under the constraints of this project.

Summary

Although 3 yrs is a relatively short time, it is significant and encouraging that environmental conditions at the cold weather sites caused no measurable or observable deterioration of sulfur foams and coatings. These tests are continuing, as three sets of specimens remain at all locations, and we are considering extending the time sequence for removing these specimens.

Sulfur-Coated Block Specimens

Through Block

Without Bactericide (psi)			With Bactericide (psi)		
1st yr	2nd yr	3rd yr	1st yr	2nd yr	3rd yr
520	590	550			
			575	535	540
650	710	720			
			590	630	560
675	775	745			
			615	735	885
555	665	830			
			615	815	780

Acknowledgments

The authors wish to acknowledge contributions and assistance from the Alberta Department of Highways, the Colorado Department of Highways, Texasgulf, Inc., and The Sulphur Institute.

Literature Cited

1. Chamberlain, D. L., Jr., "Oxidation of Sulfur," Stanford Research Institute, Menlo Park, Calif., Aug. 31, 1962.
2. Burns, G. R., Oxidation of Sulphur in Soils, Technical Bulletin No. 13, The Sulphur Institute, Jan. 1967.
3. Duecker, W. W., Estep, J. W., Mayberry, M. G., Schwab, J. W., "Studies of Properties of Sulfur Jointing Compounds," *Amer. Water Works Ass. J.* (1948) **40**, (7), 715–728.
4. Dale, J. M., Ludwig, A. C., "Cold Regions Applications for Sulfur Foams," U.S. Army Cold Regions Research and Engineering Laboratories, No. **DACA89-71-C-0024**, Feb. 1972.
5. Frederick, L. R., Starkey, R. L., "Bacterial Oxidation of Sulfur in Pipe Sealing Mixtures," *Amer. Water Works Ass. J.* (1948) **40**, 729–736.

RECEIVED May 1, 1974

11

Polyphenylene Sulfide—A New Item of Commerce

R. VERNON JONES[1] and H. WAYNE HILL, JR.
Research and Development Department, Phillips Petroleum Co., Bartlesville, Okla. 74004

> Ryton Polyphenylene Sulfide is a new commercial plastic which is characterized by good thermal stability, retention of mechanical properties at elevated temperatures, excellent chemical resistance, a high level of mechanical properties, and an affinity for a variety of fillers. It is produced from sodium sulfide and dichlorobenzene. Its unusual combination of properties suggests applications in a variety of molded parts such as non-lubricated bearings, seals, pistons, impellers, pump vanes, and electronic components. Tough coatings of polyphenylene sulfide can be applied to metals or ceramics by a variety of techniques and are used as protective, corrosion-resistant coatings in the chemical and petroleum industries. Incorporation of small amounts of polytetrafluoroethylene provides excellent non-stick properties in both cookware and industrial applications.

Traditionally, sulfur is a low cost commodity chemical used mostly in the production of sulfuric acid and other fundamental building blocks of the chemical industry. Since these uses are typified by large volume, low cost considerations, many chemists think of sulfur and sulfur compounds only in these terms. However, with ingenuity and inventiveness, high quality products which will command a much higher price and are of extreme value in the marketplace can be produced from sulfur or some of its most common compounds.

For example, large quantities of sodium sulfide have been used for many years in the leather industry to remove hair from animal hides before tanning. Large volumes of another common chemical, p-dichloro-

[1] Deceased.

benzene, have been used for many years as moth balls. However, these two common commodity chemicals react with each other under the proper reaction conditions to form a high quality polymer. This process for producing polyphenylene sulfide (PPS) is an excellent example of how chemical ingenuity can be used to convert two low cost, abundant chemicals into a premium quality plastic which can command a much higher price than either of the starting materials. This unique polymer contains about 29% sulfur and thus is a good example of high quality, sulfur-rich product.

Preparation of Polyphenylene Sulfide

A phenylene sulfide-type polymer was prepared in 1948 by Macallum (1), a Canadian working in his private laboratory. This early Macallum process involved the reaction of sulfur, sodium carbonate, and dichlorobenzene in a sealed vessel at 275–300°C.

$$Cl-\underset{}{\bigcirc}-Cl + S + Na_2CO_3 \longrightarrow \left[-\underset{}{\bigcirc}-S_x-\right]_n$$

Macallum reported that polymers prepared in this manner generally contained more than one sulfur atom per repeat unit (x in the range 1.0–1.3) (2). In addition the polymerization reaction was highly exothermic and difficult to control even on a small scale (3). Later Lenz and co-workers at Dow reported another synthesis of PPS (4, 5, 6) based on a nucleophilic substitution reaction involving the self-condensation of materials such as copper p-bromothiophenoxide. The reaction was carried out at 200–250°C under nitrogen in the solid state or in the presence of a reaction medium such as pyridine. It was quite difficult to remove the by-product, copper bromide, from polymers made by this process (7). These and other methods of polymerization have been reviewed by Smith (8). Polyphenylene sulfide resins have been described more recently by Short and Hill (9).

The process developed by Edmonds and Hill (10) in the laboratories of Phillips Petroleum Co. represents a significant departure from prior investigations and makes it possible to prepare PPS from p-dichlorobenzene and sodium sulfide by reaction in a polar solvent. (see next page).

This unique polymerization process is the basis for three significant developments: 1. the preparation of a polymer containing only one sulfur

$$\text{Cl}-\underset{}{\underset{}{\bigcirc}}-\text{Cl} + \text{Na}_2\text{S} \xrightarrow[\text{Solvent}]{\text{Heat}} \left(-\underset{}{\underset{}{\bigcirc}}-\text{S}-\right)_n + 2\ \text{NaCl}$$

atom per aromatic unit has been achieved in contrast to the variable and higher sulfur content of some of the early polymers, 2. a practical process has been discovered for producing poly-p-phenylene sulfide on a commercial basis, and 3. sufficient polymer has been available to develop practical applications for this unique material. Several grades of PPS are now manufactured and marketed by Phillips Petroleum Co. under the tradename Ryton Polyphenylene Sulfide.

Polymer Properties

In the above commercial process, PPS is produced as a fine off-white powder. X-ray diffraction studies (*11*) indicate that the polymer is highly crystalline with a crystalline melting point determined by differential thermal analysis of about 285°C. A precise characterization of the polymer is complicated by its insolubility in most solvents. At elevated temperatures, however, it is soluble to a limited extent in some aromatic and chlorinated aromatic solvents and in some heterocyclic compounds. The inherent viscosity, measured at 206°C in 1-chloronaphthalene is typically 0.16, indicating moderate molecular weight.

Polymer Curing. One of the most important characteristics of PPS is the metamorphosis caused by heating. When heated to a sufficiently high temperature (375–400°C) in air, the polymer melts to a relatively fluid liquid. On continued heating the melt becomes progressively more viscous and eventually gels and solidifies. This solid, "cured" polymer is believed to be crosslinked because it is insoluble in all solvents tested, even at elevated temperatures. Differential thermal analysis (DTA) of the cured polymer indicates either a reduction in crystallinity or its complete absence, depending on the extent of curing involved. Figure 1 compares the DTA curves of an uncured and a cured sample of PPS in a nitrogen atmosphere to suppress curing. Analysis of the uncured sample (Sample A) indicates a sharp exotherm at 125°C caused by pre-melt crystallization during heat-up and a crystalline melting point endotherm at 285°C. Analysis of the sample which was cured by heating in air at 370°C for 4 hrs (Sample B) indicates a much smaller exotherm at 135°C

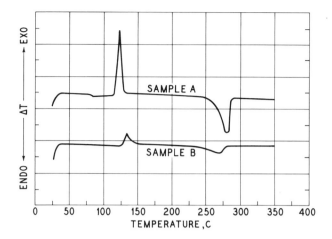

Figure 1. Differential thermal analysis of polyphenylene sulfide in nitrogen. Sample A: melted under nitrogen and quenched before DTA. Sample B: heated at 370°C in air 4 hrs and quenched before DTA. Heating rate, 10°C/minute.

caused by pre-melt crystallization and a small melting point endotherm at about 270°C. The changes in the temperature and size of both the pre-melt crystallization and melting point peaks on curing indicate a reduction in both the amount and perfection of crystallinity of the polymer.

The chemistry of this curing process involves several complex reactions. However, it is possible to describe some of the contributing

Table I. Properties of Injection-Molded Ryton PPS

Property	Unfilled PPS	40% Glass Fiber-Filled PPS
Density (g/ml)	1.34	1.64
Tensile strength (psi)	10,800	21,400
Elongation (%)	3	3
Flexural modulus (psi)	600,000	2,000,000
Flexural strength (psi)	20,000	37,000
Compressive strength (psi)	16,000	21,000
Notched izod impact strength		
25°C (ft-lb/in.)	0.3	0.8
150°C (ft-lb/in.)	1.0	1.8
Shore D hardness	86	92
Heat deflection temperature,		
at 264 psi (°C)	137	220+
Maximum recommended service		
temperature (°C)	260	260

reactions in qualitative terms. For example, a chain extension reaction involving thermal scission of carbon–sulfur bonds near the end of a polymer chain followed by formation of a new carbon–sulfur bond between two large polymer residues and between two small polymer residues can lead to an overall increase in molecular weight when the small molecules that are formed are lost by vaporization. This process is essentially an exchange reaction where m is significantly larger than n:

$$2 \left(\left(\bigcirc \right) - S \right)_m \left(\left(\bigcirc \right) - S \right)_n \xrightarrow{\Delta}$$

$$\left(\left(\bigcirc \right) - S \right)_{2m} + \left(\left(\bigcirc \right) - S \right)_{2n} \uparrow$$

$$\text{I} \qquad\qquad\qquad \text{II}$$

Thus, II is lost at the high temperatures involved in the curing process. Other reactions may also occur, for example, oxidative coupling between aromatic rings, or nucleophilic attack on an aromatic ring of one polymer chain by an end-group function of another polymer chain or by a cleaved segment derived from another polymer chain. Reactions of this type would increase molecular weight, branching, and/or crosslinking. Several of these reactions probably occur in the curing process.

In contrast to the uncured polymer, cured (heat-treated) polymers exhibit the higher melt viscosities that are desired for injection- or compression-molding and also for applying thick protective coatings of PPS. The amount of increase in melt viscosity can be controlled by the degree of curing, and several cured grades of resin are available commercially. While the uncured PPS has a moderate degree of mechanical strength, the cured resin is tough, ductile, and extremely insoluble and has outstanding chemical resistance.

Thermogravimetric analysis of PPS in nitrogen or in air indicates no appreciable weight loss below about 500°C. In air degradation is essentially complete at 700°C, but in an inert atmosphere approximately 40% of the polymer weight remains at 1000°C.

Molding Resins. Molding grades of PPS may be obtained by curing the polymer to obtain the desired melt viscosity for good moldability. These materials show an excellent affinity for reinforcing fillers such as glass fibers and asbestos. In contrast to the behavior of many plastic materials, composition of PPS containing glass fiber loadings as high as 45% can be injection-molded readily, whereas most plastic materials cannot be injection-molded at glass fiber contents greater than about 20–30%. This affinity for fillers and molding characteristics makes it possible to obtain a high level of reinforcement when PPS is used. The ease with which PPS compositions can be injection-molded parallels that of unfilled polyolefins such as polypropylene. Polyphenylene sulfide compositions flow easily into intricate mold structures and have very low mold shrinkage, thus allowing molding to close tolerances. Injection-molding is generally carried out in conventional reciprocating screw machines at a barrel temperature of 300–360°C (575–675°F) and a mold temperature of 25–200°C (75–400°F). Typical properties of unfilled and glass-filled PPS are given in Table I. Particularly noteworthy are the high flexural modulus, flexural strength, heat deflection temperature, and upper service temperature.

Polyphenylene sulfide also retains good mechanical properties at elevated temperatures, as illustrated in the plot of flexural modulus (a

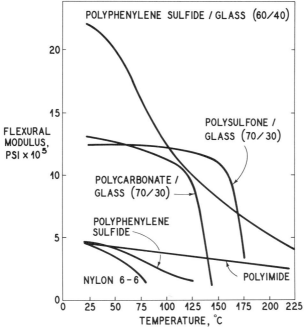

Figure 2. Temperature effect on flexural modulus

measure of stiffness) *vs.* temperature in Figure 2. Data on several conventional thermoplastics and one polyimide (Vespel SP-1) are also included in this figure.

Electrical properties of PPS compounds are summarized in Table II. The dielectric constant of 3.1 is low in comparison with other plastics. Similarly the dissipation factor is low. Dielectric strength is quite high ranging from 500 to 600 V/mil for the various compounds. Thus, both filled and unfilled PPS materials are excellent electrical insulators.

Chemical Resistance. Polyphenylene sulfide exhibits outstanding chemical resistance, because it is largely unaffected by a variety of chemicals. This behavior is summarized in Table III along with comparative data on four other well known plastics. These data were obtained by exposing tensile bars of the material to the test chemical at 93°C (200°F) for 24 hrs, measuring tensile strength after the exposure, and calculating the percent tensile strength retained after exposure. This performance among thermoplastics is surpassed only by materials such as polytetrafluoroethylene.

Flammability. Polyphenylene sulfide will not support combustion under atmospheric conditions. The oxygen index, the minimum concentration of oxygen required to maintain burning, is shown for a number of polymers (12) in Table IV. The PPS sulfide value of 44 places it among the least flammable plastics. PPS has been given a UL flammability classification of VE-O.

Molding Resin Applications. Polyphenylene sulfide is establishing itself as a basic engineering material for bearing applications and other

Table II. Electrical Properties of Polyphenylene Sulfide Compounds

Property	Unfilled PPS	40% Glass-Filled PPS
Dielectric constant, 25°C (Hz)		
10^3	3.2	3.9
10^6	3.1	3.8
10^{10}	3.1	3.6
Dielectric constant, 120°C (Hz)		
10^3	3.1	3.9
10^6	3.1	3.8
10^{10}	3.1	3.6
Dissipation factor, 25°C (Hz)		
10^3	0.0004	0.0010
10^6	0.0006	0.0013
10^{10}	0.004	0.006
Dissipation factor, 120°C (Hz)		
10^3	0.003	0.006
10^6	0.0011	0.002
10^{10}	0.02	0.007
Dielectric strength (v/mil)	585	490

Table III. Chemical Resistance[a]

Tensile Retained (%)

Chemical	Nylon 6–6	Polycarbonate	Polysulfone	Modified Polyphenylene Oxide	Polyphenylene Sulfide
37% HCl	0	0	100	100	100
10% HNO_3	0	100	100	100	96
30% H_2SO_4	0	100	100	100	100
25% H_3PO_4	0	100	100	100	100
30% NaOH	89	7	100	100	100
H_2O	66	100	100	100	100
NaOCl	44	100	100	100	84
Butyl alcohol	87	94	100	84	100
Cyclohexanol	84	74	95	27	100
Butyl amine	91	0	0	0	50
Aniline	85	0	0	0	96
2-Butanone	87	0	0	0	100
Benzaldehyde	98	0	0	0	84
Carbon tetrachloride	76	0	17	0	100
Chloroform	57	0	0	0	87
Ethyl acetate	89	0	0	0	100
Butyl phthalate	90	46	63	19	100
Butyl ether	100	61	100	0	100
p-Dioxane	96	0	0	0	88
Gasoline	80	99	100	0	100
Diesel fuel	87	100	100	36	100
Toluene	76	0	0	0	98
Benzonitrile	88	0	0	0	100
Nitrobenzene	100	0	0	0	100
Phenol	0	0	0	0	100
DMSO	84	0	0	93	100
Cresyldiphenyl phosphate	88	62	55	19	100

[a] 200°F (93°C)/24 hr.

types of anti-friction, low-wear uses. When solid lubricants such as molybdenum disulfide (MoS_2) and polytetrafluoroethylene (PTFE), etc., are incorporated, formulations with an interesting range of antifriction characteristics result. An extremely low friction coefficient and low wear rate make possible self-lubricated journal rolling elements and sliding bearing components from these Ryton PPS formulations. The need for self-lubrication becomes extremely important in those bearing applications where the bearing is so inaccessible that proper servicing is difficult or the bearing is exposed to environmental extremes that make ordinary lubricants ineffective.

Anti-friction formulations containing Ryton resin and three other available anti-friction compounds were evaluated in journal bearing test

Table IV. Flammability of Plastics

Material	Oxygen Index (%)
Poly(vinyl chloride)	47
Polyphenylene sulfide	44
Nylon 6-6	28.7
Polycarbonate	25
Polystyrene	18.3
Polyolefins	17.4
Polyacetal	16.2

configurations, and these results are summarized in Table V. Probably the most outstanding characteristic of the PPS formulation is its extremely low friction coefficient and the subsequent low bearing temperature buildup at various loads compared with the other materials. Prevention of excessive bearing temperature is most important in achieving low wear rates, particularly in plastic bearing systems where thermal conductivity is usually low and heat buildup is not dissipated nearly as efficiently as in metal bearings.

Polyphenylene sulfide molding resins offer a combination of properties which include good thermal stability, outstanding chemical resistance, low coefficient of friction, excellent electrical properties, and precision moldability. In turn, these properties lead to a variety of applications not available to many other plastics. For example, a number of pump manufacturers are using PPS compounds as sliding vanes, impeller cases, impellers, gauge guards, and seals in corrosive service involving materials such as 60% sulfuric acid, liquid ammonia, and various hydrocarbon streams.

The electrical properties of PPS and the ability to injection-mold very small parts with great precision have led to the use of a variety of connectors, coil forms, *etc.* in the electronics industry. For example, a pin cushion corrector coil used in color television sets is now being produced

Table V. Comparative Journal Bearing Data[a]

Material Designation	Radial Wear (in.)	Test Duration (hrs)	Friction Coefficient (μ)	Maximum Bearing Temp. (°F)	Maximum Shaft Temp. (°F)
PPS/MoS_2/Sb_2O_3	0.0060	162.5	0.02–0.05	155	<200
PTFE/MoS_2/fiberglass	0.0025	125	0.23–0.26	316	347
PTFE/glass/iron oxide	0.0084	117.5	0.20–0.30	340	>500
Polyimide/PTFE/graphite	0.0078	55	0.08–0.18	205	277

[a] Measurements made at 60 psi load and 1800 rpm, corresponding to a PV of 17640 (load in psi times speed in ft/min). Test shaft surface finish was 6–10 μin., root mean square.

from PPS (*13*). In another application a 10.5-in. diameter piston for a non-lubricated gas compressor has been in service at 1000 rpm for over 6 months and is performing better than the aluminum piston which it replaced. This piston was machined from a 35-lb compression-molded block of PPS.

Coatings. Polyphenylene sulfide coatings can be applied to aluminum, steel, and other materials by a variety of solventless techniques such as slurry spraying from aqueous systems, fluid bed techniques, electrostatic spraying, and powder spraying. Each of these techniques required a bake cycle at elevated temperatures to coalesce the polymer particles and to cure the polymer. Typical cure conditions are 45 min at 370°C (700°F) for coatings 1–2 mils thick. Thicker coats require somewhat longer cure times. Pigments can be added to provide a variety of colors; 33 parts titanium dioxide/100 parts PPS is a common pigment system.

Incorporation of small amounts of polytetrafluoroethylene in the PPS coating formulation provides hard, scratch-resistant release coatings (*14*). The use of release coatings of this type in food contact service is permitted by Section 121.2621 of the Food Additives Regulations.

The properties of two typical PPS coatings are summarized in Table VI. These coatings are characterized by excellent adhesion, hardness, chemical resistance, thermal stability, and flexibility.

Table VI. Properties of Polyphenylene Sulfide Coatings on Steel

Property	PPS/TiO_2 Coating (3/1)	$PPS/TiO_2/PTFE$ Coating (3/1/0.3)
Hardness, pencil	2H	2H
Mandrel bend, 180°, 3/16 in.	pass	pass
Elongation (ASTM D 522) (%)	32	32
Reverse impact (in.-lb)	160	160
Abrasion resistance, Taber mg loss/1000 rev, CS-17 wheel	50	57
Chemical resistance	excellent	excellent
Thermal stability	excellent	excellent
Color	light tan	light tan

Polyphenylene sulfide release coatings are being used in non-stick cookware. In another application, tire molds are coated to aid the release of the finished tire from the mold (*15*). When this release coating was used, more than 8000 tires were produced without cleaning the mold, whereas cleaning was required after producing only 500–600 tires when a conventional silicone mold release agent was used. In addition the need for a blemish paint was eliminated, tire rejects were reduced, and an improved surface finish resulted.

Polyphenylene sulfide coatings are also being used as corrosion-resistant, protective coatings for oil field pipe, valves, fittings, couplings, thermocouple wells, and other equipment in both the petroleum and chemical processing industries. Coated parts of this type have been operating satisfactorily for extended periods in media such as liquid ammonia, crude oil, refined hydrocarbons, brine, dilute hydrochloric and sulfuric acids, dilute caustic, and many other chemicals. In particular, PPS is providing protection when both corrosive environments and elevated temperatures are involved. Thus parts of carbon steel coated with PPS formulations are replacing parts previously fabricated from expensive alloy metals.

Summary

Polyphenylene sulfide is a unique material which is characterized by good thermal stability, retention of mechanical properties at elevated temperatures, excellent chemical resistance, a high level of mechanical properties, and an affinity for a variety of fillers. This unusual combination of properties provides application for this resin in a variety of molded parts such as non-lubricated bearings, seals, pistons, impellers, pump vanes, and electronic components. Tough coatings of PPS can be applied to metals by a variety of techniques and are being used as protective, corrosion-resistant coatings for the chemical and petroleum industries. Incorporation of small amounts of polytetrafluoroethylene provides excellent non-stick properties, and these hard release coatings are used in both cookware and industrial applications.

Polyphenylene sulfide is produced from sodium sulfide and dichlorobenzene by a novel polymerization process. This process is an excellent example of the use of chemical ingenuity in converting two low cost, abundant chemicals into a premium quality plastic which can command a much higher price than either of the starting materials. This unique polymer contains about 29% sulfur and thus is a good example of a high quality, sulfur-rich product.

Literature Cited

1. Macallum, A. D., *J. Org. Chem.* (1948) **13**, 154.
2. Macallum, A. D., U.S. Patents **2,513,188** (June 27, 1950) and **2,538,941** (Jan. 23, 1951).
3. Macallum, A. D., private communication.
4. Lenz, R. W., Carrington, W. K., *J. Polym. Sci.* (1959) **41**, 333.
5. Lenz, R. W., Handlovits, C. E., *J. Polym. Sci.* (1960) **43**, 167.
6. Lenz, R. W., Handlovits, C. E., Smith, H. A., *J. Polym. Sci.* (1962) **58**, 351.

7. Smith, H. A., Handlovits, C. E., "Phenylene Sulfide Polymers," **ASD-TRD-62-322,** Part II, pp. 18–19, Aeronautical Systems Division, Air Force Systems Command, 1962.
8. Smith, H. A., *Encycl. Polym. Sci. Technol.* (1969) **10,** 653.
9. Short, J. N., Hill, H. W., Jr., *Chemtech* (1972) **2,** 481.
10. Edmonds, J. T., Jr., Hill, H. W., Jr., U.S. Patent **3,354,129** (Nov. 21, 1967).
11. Tabor, B. J., Magre, E. P., Boon, J., *Eur. Polym. J.* (1971) **7,** 1127.
12. Goldblum, K. B., *SPE (Soc. Plast. Eng.) J.* (Feb. 1969) **25,** 50.
13. *Plast. Des. Process.* (July 1973) **13,** 21.
14. Tieszen, D. O., Edmonds, J. T., Jr., U.S. Patent **3,622,376** (Nov. 23, 1971).
15. *Plast. Des. Process.* (Sept. 1973) **13,** 76.

RECEIVED May 1, 1974

12

Chemical Investigations of Lithium-Sulfur Cells

J. R. BIRK[1] and R. K. STEUNENBERG

Chemical Engineering Division, Argonne National Laboratory, Argonne, Ill. 60439

> *High-performance lithium–sulfur cells are being developed for use in off-peak energy storage batteries in electric utility networks. The cells, which operate at 400°C, consist of a lithium electrode, a sulfur or sulfide electrode, and molten LiCl–KCl electrolyte. The chemistry of the lithium electrode is relatively straightforward. However, the electrochemical reactions at the sulfur electrode involve the formation of several intermediate species that are sufficiently soluble in the electrolyte to limit the lifetime and capacity of the cells. Although this effect can be decreased with soluble additives such as arsenic or selenium in the sulfur, a more promising solution appears to be the use of metal sulfides, rather than sulfur, as the active material in the positive electrode.*

High-performance lithium–sulfur secondary batteries are being developed for use in electric automobiles and for off-peak energy storage in electric utility systems. These applications impose severe performance requirements that cannot be met by present batteries. For the electric automobile, the battery must have a minimum specific energy of ∼200 W-hr/kg, a specific power of at least 200 W/kg, and a lifetime of 3–5 yrs. The projected performance requirements for a battery for off-peak energy storage are a maximum cost of $12–15/kW-hr, a specific power of ∼50 W/kg, and a minimum lifetime of 5–10 yrs.

The development of lithium–sulfur batteries began in 1967 as a small basic research effort at Argonne National Laboratory (1). Since then,

[1] Present address: Electric Power Research Institute, P. O. Box 10412, Palo Alto, Calif. 94304.

significant progress has been made both at Argonne and at other laboratories, including Atomics International and General Motors. At present, the Argonne program is entering the early hardware stage, with the construction and testing of a full-scale sealed cell of 120 A-hr (156 W-hr) capacity and a specific energy of 104 W-hr/kg (2). Recent developments in both hardware and cell life indicate that a lithium–sulfur battery system capable of meeting the above requirements may be forthcoming within 5 yrs. Subsequent commercial development and production of the batteries may require another few years, thereby placing large-scale practical application in the early 1980s. The general aspects of the development work on lithium–sulfur batteries have been described in various technical papers and progress reports. This particular paper is concerned mainly with the chemical and electrochemical processes occurring in the cells. In addition, recommendations that have resulted in improved cell performance, some of which were derived as a result of this investigation, will be discussed and evaluated.

Lithium and sulfur are promising active electrode materials for batteries because of their low equivalent weight, low cost, and suitable electrochemical properties. The major question in the use of these active electrode materials is whether the electrodes will be able to provide sustained high performance. Until recently, capacity retention over an extended period has been difficult to achieve.

Chemistry of the Lithium Electrode

The capacity loss from the lithium electrode has resulted mainly from dewetting of metallic substrate materials, such as nickel and stainless steel Feltmetal and reaction with the electrolyte, the LiCl–KCl eutectic, to form potassium vapor:

$$\text{Li(l)} + \text{KCl(l)} \rightarrow \text{K(g)} + \text{LiCl(l)} \quad (1)$$

This reaction results in an equilibrium potassium vapor pressure (calculated from thermodynamic data) of 0.714 torr above the LiCl–KCl eutectic at 427°C (700°K). Metallic lithium is rapidly lost, by Reaction 1, from the lithium electrode in open cells exposed to an inert atmosphere of helium (3). However, this reaction has not been evident in hermetically sealed cells.

Lithium is soluble in the LiCl–KCl electrolyte at about 0.13 mole % at 400°C, based on an extrapolation of the data from the literature (4, 5). This solubility does not lead to significant lithium loss. However, the electronic conductivity of the lithium-saturated salt at 450°C is estimated (6, 7) to be 5×10^{-3} ohm^{-1} cm^{-1}, which could lead to extensive self-discharge. Fortunately, self-discharge rates of lithium–chlorine (8) and lithium–sulfur cells are much lower than that predicted from the electronic conductivity of the electrolyte saturated with lithium (8). It is possible that oxidation of dissolved lithium at the sulfur electrode depletes the lithium concentration at the electrode surface, thereby causing a break in the electronic circuit within the electrolyte.

The dewetting problem of the lithium electrode is now under intensive investigation. General Motors has found that high-temperature hydrogen treatment of the nickel Feltmetal substrate improves lithium retention (9). In this process, metal-oxide films are removed from the substrate metal and the lithium and substrate are in more intimate contact.

In this laboratory, solid lithium–aluminum alloys and solutions of lithium and metal additives are being tested. This latter approach takes advantage of the fact that solubility and wettability are related (10) and that certain metal additives are somewhat soluble in lithium at 400°C. As a result, good electrical and physical contact can be maintained between the lithium and metal additive and between the metal additive and substrate metal.

Chemistry of the Sulfur Electrode

The capacity loss of the sulfur electrode can be attributed to sulfur vaporization (the vapor pressure of sulfur at 400°C is 410 torr), migration or dispersion of insoluble sulfur-containing phases from the electrode, solubilization of sulfur-containing species in the electrolyte, and/or inactivation of sulfur within the electrode compartment. Since neither the mechanism of the cell reaction nor the mechanism of sulfur loss was understood, a study of sulfur electrode chemistry and electrochemistry was made. It was expected that information gained from these studies would lead to improved performance and lifetimes of lithium–sulfur cells.

Cyclic Voltametric Studies. The major investigative tool used to study the chemistry and electrochemistry of the sulfur electrode was cyclic voltammetry. The working solution contained Li_2S (≤ 0.065 wt % which is equivalent to $0.024M$) as the initial electroactive solute and the LiCl–KCl eutectic as the solvent. Elemental sulfur was not chosen as the starting material because it is virtually insoluble in the LiCl–KCl eutectic and also because it is very volatile. It was expected, therefore, that a homogeneous solution of sulfur could not be prepared as is possible with a dilute solution of Li_2S in the LiCl–KCl eutectic. Polysulfide was not selected as the initial reactant because it disproportionates into several other electroactive species. This, in turn, would limit the ability to identify the reactions occurring at the working electrode. The investigation involved the electrolyte (the LiCl–KCl eutectic) rather than the molten sulfur portion of the positive electrode, because the chemistry involved in the electrolyte will often determine the long-term performance of a cell. Previous electrochemical studies (11, 12, 13, 14, 15) of the reduction of sulfur in the LiCl–KCl eutectic have not definitely identified the reaction mechanism involved in the electrochemical reduction of sulfur.

Previous Work. Bodewig and Plambeck (11) suggested that sulfur reduction involves a single step:

$$S + 2e^- \rightarrow S^{2-} \tag{2}$$

In arriving at this reaction, the authors assumed the identity of the product (Li_2S), which they generated electrochemically, in order to prove their overall reaction. In addition, the concentrations of Li_2S, which they indicated were generated in the LiCl–KCl eutectic, were much above the solubility limit (0.029M at 425°C) as determined by Liu et al. (16).

Bernard et al. (12, 13) used chronopotentiometry and spectrophotometry to demonstrate that two intermediates, a polysulfide (S_n^{2-}) and a supersulfide (S_n^-), are involved in the sulfide oxidation. A definitive reaction mechanism, however, could not be proposed because the authors stated (12) that their reaction was spontaneous and indicated (13) that they could not identify the oxidizing agent.

Recent cyclic voltammetric studies by Kennedy and Adamo (14) and by Cleaver et al. (15) indicate that sulfur reduction involves two steps. These authors observed two peaks on both anodic and cathodic scans which they concluded were due to the two following steps:

$$nS \overset{e^-}{\rightleftarrows} S_n^- \overset{e^-}{\rightleftarrows} S_n^{2-} \tag{3}$$

Neither group of workers indicated or mentioned a further reduction (e.g., Reaction 4):

$$S_n^{2-} + (2n-2)e^- \rightleftarrows nS^{2-} \tag{4}$$

The mechanism proposed by Kennedy and Adamo (14) was based on the "similarity of the results obtained in DMSO (dimethylsulfoxide)" where the presence of S_8^- and S_8^{2-} was hypothesized (17). However, it is questionable whether results obtained with an organic solvent at 25°C can be compared with those obtained with an inorganic molten salt at 420°C. In addition, the proposed mechanism does not appear to satisfy all the data obtained by Cleaver et al. (15).

It was apparent that there was a lack of conclusive data to establish a mechanism for the reduction of sulfur to sulfide in the LiCl–KCl eutectic. In addition, this type of information was needed so that further improve-

Table I. Experimental Conditions

Cell	alumina beaker, 5.0 cm id, 8.3 cm tall
Electrodes	
Working	0.32-cm diam. National Spectroscopic carbon
Reference	stainless steel Feltmetal containing liquid lithium, 1.27 cm diam. × 0.63 cm thick
Counter	0.63 cm diam. National Spectroscopic carbon
Working electrode area	0.24 cm²
Electrolyte	polarographic grade LiCl–KCl eutectic from Anderson Physics Laboratories, ca 100 g
Cell compartment separators	1.0-cm diam. medium porosity quartz frits
Initial electroactive species	lithium sulfide (Foote Mineral), stated purity 97%
Working solution	Li_2S–LiCl–KCl, max. concentrations of Li_2S (0.05 and 0.065 wt %) as prepared, other concentrations computed using first anodic peak height
Cell resistance	10–13 ohms (counter to working electrode)

ments in sulfur electrode performance could be attained in lithium–sulfur cells. Thus, a laboratory study was undertaken to provide more data to investigate further the mechanism involved in the electrochemical reduc-

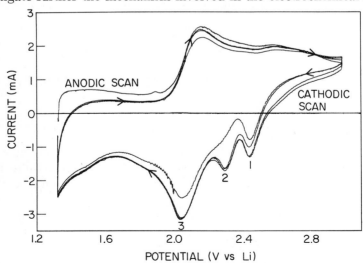

Figure 1. Cyclic voltammetry of Li_2S–LiCl–KCl at 410°C. Concentration $Li_2S = 0.006$ wt %. Scan rate = 66 mV/sec.

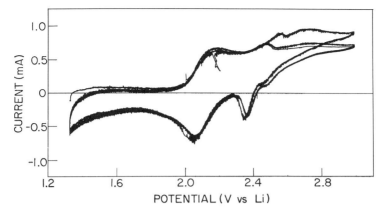

Figure 2. Cyclic voltammetry of Li_2S–$LiCl$–KCl at 410°C. Concentration $Li_2S = 0.006$ wt %. Scan rate = 6.6 mV/sec.

tion of sulfur. The experimental conditions used in this investigation are given in Table I.

Cyclic Voltammetric Results and Discussion. Essentially three different types of studies were conducted in which the cyclic voltammograms were obtained as a function of potential scan rate (Figures 1–3), Li_2S concentration (Figures 1 and 4), and temperature (Figures 5 and 6). The data show that the first peak on cathodic scan decreases and the second one increases at higher sulfide concentrations, at increased temperatures, and at longer reaction times (*i.e.*, slower scan rates). The data shown in Figures 1–6 are typical of those taken during this investigation.

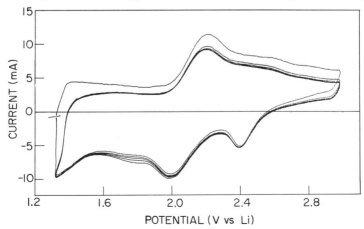

Figure 3. Cyclic voltammetry of Li_2S–$LiCl$–KCl at 410°C. Concentration $Li_2S = 0.006$ wt %. Scan rate = 660 mV/sec.

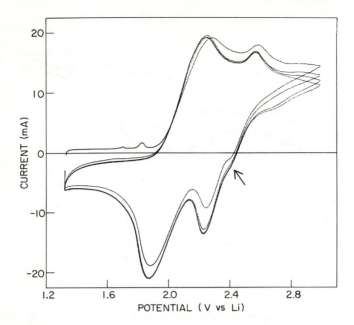

Figure 4. Cyclic voltammetry of Li_2S–LiCl–KCl at 410°C. Concentration $Li_2S = 0.065$ wt %. Scan rate $= 66$ mV/sec.

Five different Li_2S concentrations ranging from about 0.002 to about 0.065 wt % run at several widely different scan rates (6.6–660 mV/sec) confirm the behavior depicted in Figures 1–4. In addition, temperature studies between 410°C (Figure 5) and 462°C (Figure 6) show the gradual decrease in the first cathodic peak and a gradual increase in the second cathodic peak.

These results suggest that, in addition to the three electrochemical reactions that are observed during cathodic scan, there is at least one chemical reaction. The chemical reaction appears to involve sulfide and the species responsible for the first peak observed during the cathodic scan. The product of the reaction appears to be the species responsible for the second cathodic peak. An increase in reaction time, temperature, and/or sulfide concentration would be expected to enhance such a chemical reaction, thus causing a decrease in the first peak and an increase in the second peak as was observed. Since the species responsible for the second cathodic peak is formed only by chemical reaction and not by electrochemical reaction, no second peak is observed (Figure 3) at rapid scan rates where the extent of chemical reaction appears negligible.

The third cathodic peak and the remaining cathodic scan are unaffected by changes observed in the first two cathodic peaks (*see*, for example, Figures 5 and 6). Therefore, the same species must be formed

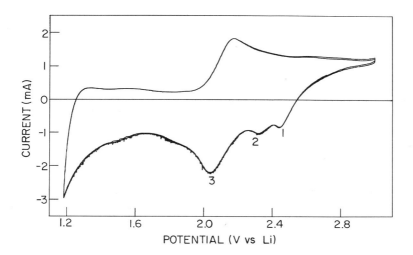

Figure 5. Cyclic voltammetry of Li_2S–$LiCl$–KCl at 410°C. Concentration $Li_2S = 0.01$ wt %. Scan rate = 36 mV/sec.

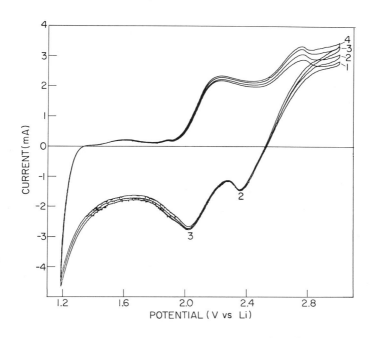

Figure 6. Cyclic voltammetry of Li_2S–$LiCl$–KCl at 462°C (cycle 1) to 478°C (cycle 4). Concentration $Li_2S = 0.01$ wt %. Scan rate = 36 mV/sec.

from the electrochemical reactions responsible for the first and second cathodic peaks, and the reduction of this species is responsible for the third peak.

The cathodic current observed at the left-hand portion of the voltammograms (particularly significant in Figures 5 and 6) was observed with a carbon, but not with a tungsten working electrode. Although this result is not fully understood, the current may result from the reduction of absorbed oxygen (which could form carbon dioxide) or from the formation of lithium carbide. Free energy calculations show that the standard potential for lithium carbide formation at 427°C is 0.2 V *vs.* lithium, and for carbon dioxide reduction to carbon monoxide is 1.3 V.

On anodic scan, none of the cyclic voltammograms has shown more than two peaks. The second anodic peak is very small relative to the first except at elevated temperatures or at relatively high sulfide concentrations. The cyclic voltammetric results suggest that the oxidation mechanism occurring subsequent to the first oxidation reaction is quite complex. Comments concerning these oxidation reactions are given following the discussion of the overall reaction mechanism which is based largely on the results from the cathodic portion of the cyclic voltammograms.

The cyclic voltammograms obtained during this investigation are similar in many respects to those obtained by Kennedy and Adamo (*14*) and by Cleaver *et al.* (*15*). Thus, for example, the first anodic peak and the last cathodic peak appear to represent a reversible reaction with a peak potential separation of about 0.10 V. This separation of peak potentials corresponds to a value of between one and two for the number of electrons transferred. More likely this could represent an unsymmetrical reaction involving a different number of electrons transferred by the reactive species on charge and that on discharge. In addition, the peak potential separation (0.4 V) between the first cathodic and last cathodic peaks and that (0.3 V) between the first cathodic and first anodic peaks in this investigation and that of Cleaver *et al.* (*15*) were identical.

The major difference in the results appears to be the presence during the cathodic scan of three reduction peaks in this work instead of the two reported by Kennedy and Adamo (*14*) and Cleaver *et al.* (*15*). Careful inspection of the data obtained and the approach used by these workers (*14, 15*) suggests that the middle (second) cathodic peak observed in this investigation was not present in the voltammograms of the previous investigators. The presence of Li_2S in the bulk electrolyte was necessary before this peak could be observed and, even then, it could only be seen at specific scan rates and temperatures. It is, therefore, understandable why this peak was not observed by the previous investigators who used sulfur (*14*) and $Na_2S_{2.2}$ (*15*) instead of Li_2S as the initial electroactive species. The presence of this peak necessitates a re-evaluation of the conclusions suggested in the earlier work (14, 15).

The first cathodic peak most probably results from the sulfur reduction. This is suggested by the similar results of this investigation and previous electrochemical studies (14) where the initial electroactive species was sulfur and by results from lithium–sulfur cell tests where sulfur is known to form from the electrochemical oxidation of Li_2S.

Comparisons of the first and subsequent cycles show that the anodic peaks do not change significantly with time, suggesting that the initial electroactive species, sulfide, is reformed during the reduction process. Thus, the last cathodic and the first anodic peaks must involve the formation and oxidation, respectively, of sulfide.

Based on the results of a literature review, the intermediates responsible for the second and third cathodic peaks appear to be a supersulfide (S_n^-), and polysulfide (S_n^{2-}), respectively. The formation of supersulfide ions (S_2^- and/or S_3^-) in the LiCl–KCl eutectic and in KSCN has been suggested by several authors (18, 19, 20, 21, 22). Gruen et al. (18) and Vogel et al. (19) observed by absorption spectrophotometry the formation of these species from a reaction between dissolved sulfide in the LiCl–KCl eutectic and sulfur vapor at 400–800°C. They demonstrated that both S_2^- (hypersulfide) and S_3^- (not yet named) exist in the molten salt (18, 19). However, Vogel et al. (19) suggest that the hypersulfide is the more stable of the two species. For example, the equilibrium constant for the formation of S_2^- is 300 times greater than that for S_3^- at 500°C (19). At less than 500°C, dimerization of the supersulfides (which, having a free unpaired electron, are radicals) may become important (21). The products are polysulfides which are somewhat soluble but are also unstable (particularly when n is greater than about 2) in the LiCl–KCl eutectic at 400°C (15).

In the lithium–sulfur system alone (i.e., no electrolyte), these higher polysulfides are unstable because only Li_2S_2 is known (23). However, recent phase diagram studies appear to suggest that Li_2S_3 or LiS_2 (or its dimer, Li_2S_4) may be a stable species above 360°C. These studies demonstrate that there is a liquid phase containing 68.4 mole % sulfur and 31.6 mole % Li_2S, according to Cunningham et al. (24) and 73.7 mole % Li_2S according to Sharma (25). Independent of the species that may exist above 360°C, the only stable compound below this temperature is Li_2S. In other words, the phase corresponding to a composition of about LiS_2 decomposes to sulfur and Li_2S at temperatures less than 360°C. Thus, x-ray diffraction studies on solidified samples from the sulfur electrode do not provide a valid identification of the intermediate species between sulfur and sulfide that may be present in the liquid phase at the cell operating temperature.

Proposed Mechanism. On the basis of the results from cyclic voltammetry and those brought out by the literature review, the following mechanism is proposed for the electrochemical reduction of sulfur in the LiCl–KCl eutectic:

$$3\ S + S^{2-} \rightleftarrows 2\ S_2^- \rightleftarrows S_4^{2-} \tag{5}$$

$$2\ S + 2\ e^- \rightleftarrows S_2^{2-} \tag{6}$$

$$S_2^-\ (\text{or}\ S_4^{2-}) + 1\ e^-\ (\text{or}\ 2\ e^-) \rightleftarrows S_2^{2-} \tag{7}$$

$$S_2^{2-} + 2\ e^- \rightleftarrows 2\ S^{2-} \tag{8}$$

These reactions are shown on a typical cyclic voltammogram in Figure 7. The data suggest that hypersulfide is only formed chemically (Reaction 5) and not by an electrochemical reaction. In Reaction 6, it is possible that higher sulfides could be formed. However, S_2^- was indicated as the product because it is the most stable polysulfide.

The anodic electrochemical reactions which can occur following the reverse of Reaction 8 are: the oxidation of sulfide, present in the bulk electrolyte, to polysulfide; the oxidation of polysulfide to hypersulfide and/or sulfur; and the oxidation of hypersulfide (which could form either

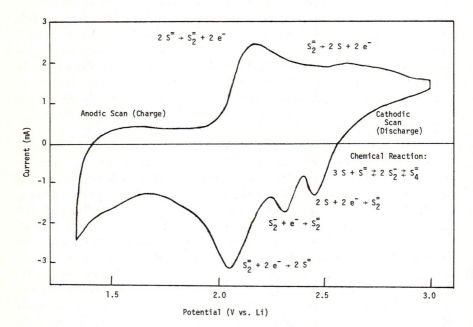

Figure 7. Cyclic voltammetry of Li_2S–$LiCl$–KCl at 410°C. Concentration $Li_2S = 0.006\ wt\ \%$. Scan rate = 66 mV/sec.

electrochemically or chemically) to sulfur. Depending on the conditions, it appears that various combinations of the above reactions can occur.

The data suggest that sulfur formation on the electrode surface inhibits further oxidation of sulfide or polysulfide because extensive oxidation is absent after the first anodic peak and a sharp second anodic peak is absent under conditions where the chemical reaction is slow such as lower temperatures and sulfide concentrations (see Figures 1, 3, and 5). However, under conditions conducive to Reaction 7, there is a significant increase in the extent of oxidation (see Figures 1 and 4, and Figures 5 and 6). These data suggest that the sulfur film can be removed by Reaction 7 to form hypersulfide and/or its dimer. With the film removed, the hypersulfide can be readily oxidized to sulfur. Thus, the chemical reaction appears to assist in the oxidation process.

Having such a complex mechanism take place at the sulfur electrode is not necessarily harmful. However, evidence suggests that, in this case, the intermediates formed are somewhat soluble in the LiCl–KCl electrolyte, whereas sulfide (15, 16) and sulfur (13) are only sparingly soluble. This was demonstrated, for example, by the comparative solubilities of the intermediates and the sulfur and sulfide species in the LiBr–RbBr eutectic salt at about 365°C (26). The solubility of sulfur in the eutectic was found to be 0.3 mole % (0.08 wt %) and that of the sulfide to be 1.5 mole % (0.6 wt %), whereas the solubility of a mixture of the two, where intermediate formation could occur, was 11 mole % (3.6 wt %). Being soluble in the electrolyte, these intermediates can migrate to the lithium electrode and react to form solid Li_2S. Migration of the sulfide back to the sulfur electrode would be limited by its low solubility in the electrolyte. Based upon these considerations, it is not surprising that examinations of lithium–sulfur cells have shown deposits of Li_2S around the lithium electrode. The major implication of these cell studies and electrochemical investigations is that the intermediates formed during the cell reaction probably cause sulfur loss from the sulfur electrode. As a result, to retain sulfur, intermediate formation must be prevented. What is desired, therefore, is to promote Reactions 9 and 10.

$$2S_2^- \rightarrow 3S + S^{2-} \qquad (9)$$

$$S_2^2 \rightarrow S^{2-} + S \qquad (10)$$

These reactions as written, however, appear to be thermodynamically unfavorable. Therefore, stabilization of the products, S or S^{2-}, is one possible way to enhance decomposition of the intermediates. This can be accomplished by using additives with the sulfur or by using sulfur compounds as the active electrode material. In addition, using compounds and additives not only diminishes sulfur loss by solubilization but also inhibits potential losses by migration or dispersion and by vaporization.

Use of Additives in Sulfur Electrodes. The use of additives to diminish sulfur loss and, thus, to retain cell capacity, has been investigated at Argonne for several years. Thus far, the most successful cells using sulfur-additive mixtures have incorporated selenium or arsenic as the additive. Cells using these additives have performed much better than those using no additives. However, it appears that these particular additives, at least, do not lower the activity of the sulfur sufficiently for high performance of the lithium–sulfur cell over long periods.

Atomics International has used several sulfur compounds as active sulfur-electrode materials (27). In this laboratory, FeS_2, FeS, Sb_2S_3, As_2S_2, As_2S_3, and P_4S_{10} have been tested. FeS_2 has been the most intensively investigated and has resulted in long-term capacity retention.

The use of compounds to replace sulfur in the positive electrode necessitates a terminology change with respect to the positive electrode, which can no longer be considered a sulfur electrode. In the case of metal sulfide compounds, the sulfur will not be an electroactive species (except in such compounds as FeS_2). The only function of the sulfur is to retain the reactant (*i.e.*, the metal) and the product (*i.e.*, Li_2S) within the electrode cavity.

The Properties of Sulfur-Additive Mixtures

Vapor Pressure of Sulfur-Additive Mixtures. Sulfur vaporization can be a significant source of sulfur loss from the sulfur electrode, particularly on cell start-up, because sulfur is highly volatile at operating temperatures. Steps to decrease sulfur loss during start-up have included using sulfur-soluble additives and starting the cell from a discharged state (*i.e.*, using Li_2S as the starting material in the sulfur electrode). Additives have lowered the sulfur vapor pressure quite successfully because a small amount of any soluble additive would be expected to lower the vapor pressure of sulfur as shown by Equation 11:

$$P_s = (X_s)^n P_s^\circ \tag{11}$$

This is Raoult's Law, modified for the case of polymerization in the vapor phase, where P_s is the partial pressure of sulfur above a solution containing sulfur, P_s° is the vapor pressure of pure sulfur, X_s is the mole fraction of sulfur in the solution, and n is the number of sulfur atoms in a molecule of vapor [*i.e.*, $nS(1) \rightleftarrows S_n(g)$]. At 444°C and at 1 atm, the average value of n is 6.50 (28).

Using this value for n, it can be shown that 10 at. % soluble additive lowers the sulfur vapor pressure by about 50% and 20 at. % lowers the vapor pressure by about 75%. To confirm that Equation 11 applies to sulfur–arsenic and sulfur–selenium mixtures, calculated and observed boiling points of these mixtures were compared as shown in Table II. There is little deviation between observed and theoretical boiling point elevations for dilute solutions of arsenic and selenium in sulfur, except for the mixture of 15.5 at. % selenium in sulfur where the deviation is 42°C. The close relationship between the two values shows that the sulfur behaves nearly ideally with dilute solutions of arsenic (<20 at. %) and selenium (<5–10 at. %). It follows, therefore, that Equation 11, which also assumes ideal behavior for the solvent, also applies for these dilute mixtures of selenium and arsenic in sulfur.

Table II. **Boiling Point Elevation for Sulfur–Arsenic and Sulfur–Selenium Mixtures**

Composition (at % solute)	Boiling Point Elevation (°C)	
	Observed[a]	Theoretical[b]
0	0	0
3.94	17 (Se)	18
10.0	48 (As)	47
15.5	41 (Se)	83
20.0	105 (As)	115

[a] The observed values for S–As mixtures were taken from the phase diagram (29), and those for S–Se mixtures were calculated using equations relating total vapor pressure observed above various S–Se mixtures to temperature (30).

[b] The theoretical values were calculated from the boiling point elevation equation, derived for a system that involves polymers in the gas phase.

$$\Delta T_B = \frac{nR\, T_o T_B\, (X_B + 1/2\, X_B^2 + 1/3\, X_B^3 + \ldots)}{\Delta H_{vap}}$$

where: T_B is the boiling point of the sulfur-solute mixture; T_o is the boiling point of sulfur; ΔT_B is the boiling point elevation; X_B is the mole fraction of solute; R is the gas constant; ΔH_{vap} is the heat of vaporization per mole of vapor (15,1000 cal/mole for sulfur) obtained from R times the slope of $\ln P$ vs. $1/T$; and n is defined as in Equation 11. It is assumed to be 6.50 over the temperature range investigated. Also, it is assumed that the value for n is unaffected by the presence of small amounts of arsenic in the gas phase.

Reactions of Compounds and Sulfur-Additive Mixtures. When additives and compounds are used to retain sulfur in the positive electrode there are other possible reactions, in addition to the usual electrochemical reactions, that can take place at the positive electrode. These include electrochemical oxidation and reduction of the additive (or compound) and interaction of the additive with the electrolyte. In addition, there is the possibility of ternary compound formation from a reaction between the discharge product, Li_2S, and sulfur compounds. Because the arsenic–sulfur system has undergone the most intensive investigation for use in lithium–sulfur cells, examples of the aforementioned reactions using arsenic are shown below:

$$3\ Cl^- + As \rightarrow AsCl_3\ (g) + 3\ e^- \qquad (12)$$

$$3\ Li + As \rightarrow Li_3As \qquad (13)$$

$$Li_2S + As_2S_3 \rightarrow 2\ LiAsS_2 \qquad (14)$$

An example of an interaction between the electrolyte and a sulfur compound is shown for the B_2S_3 system in LiF–LiCl–KCl:

$$B_2S_3 + 6\ LiF \rightarrow 2\ BF_3\ (g) + 3\ Li_2S \qquad (15)$$

Reaction 12 is undesired because it involves loss of the additive as a gas. The potential at which this reaction may occur can be calculated from free energy data. For arsenic, the potential of Reaction 12 is 2.90 V *vs.* lithium. This was confirmed in two tests, one a lithium–sulfur–arsenic cell, the other a lithium–arsenic cell. In the cell containing the sulfur–10 at. % arsenic mixture, a plateau was observed upon charging at 2.95–3.15 V (IR-free with a current density of 20–60 mA/cm^2). With the lithium–arsenic cell, the plateau on oxidation was at 2.87 V (IR-free at about 70 mA/cm^2).

Reaction 13 may be advantageous during cell operation. It can result in additional capacity of the sulfur electrode. This sort of reaction appears to be evident for antimony in a lithium–Sb_2S_3 cell at about 0.9–1.2 V. For arsenic, the open-circuit potential of a lithium–arsenic cell was observed to be 1.12 V. Most of the discharge capacity in this cell was in the range of 0.85 to 1.05 V. It is evident that the discharge potential is impractically low to benefit significantly cells containing antimony or arsenic.

Reaction 14 is an example of a reaction that can occur between active sulfur compounds and the reaction product, Li_2S. The possibility of this type of reaction was investigated by heating equimolar amounts of Li_2S and arsenic trisulfide in an evacuated quartz ampoule at 385°C and then at 480°C for a total of 5 hrs. X-ray diffraction showed that little, if any, of the reactants were present and that a compound, possibly $LiAsS_2$, had been formed. It is likely that compounds of this type could stabilize the sulfur and sulfide, which would assist in decreasing the extent of intermediate (S_2^- and S_2^{2-}) formation by Reactions 9 and 10. However, the ternary compounds must be insoluble in the electrolyte to prevent loss from the electrode cavity.

Reaction 15 appeared to take place in a cell using B_2S_3 as the active electrode material. Gas was evolved for a long time without electrochemical operation of the cell. Free energy estimates indicate that Reaction 15 was a likely cause of this gas formation.

Many reactions of the general nature shown here are possible with each additive or compound. However, free energy calculations can be used to determine whether these reactions can occur.

Conclusions

Fundamental chemical investigations are being used to not only study the interesting chemistry of the lithium–sulfur cell but also to develop solutions to problems encountered in cell operation. Thus, chemical investigations of the sulfur electrode have suggested that elemental sulfur in combination with the electrode reaction product, Li_2S, cannot be contained in the positive electrode compartment for a long time. The data suggest excessive, irreversible solubilization losses in the electrolyte for such an electrode. Furthermore, these investigations suggest that the use of sulfur compounds and/or possibly sulfur-additive mixtures as the active electrode material is a suitable means of preventing sulfur loss. Although the new sulfide electrode concepts being investigated do not involve elemental sulfur, the impact of this change from sulfur to sulfide electrodes to the sulfur industry is negligible because even widespread use of lithium–sulfur batteries would involve only a relatively small amount of sulfur.

Acknowledgments

The authors thank Leslie Burris, Donald Webster, and Paul Nelson, management of the Chemical Engineering Division, Argonne National Laboratory, for providing a scientific atmosphere conducive for independent and fundamental studies to supplement the engineering aspects of battery development. The authors also thank Donald Vissers for his valuable comments concerning the chemistry of lithium electrode development. In addition, the authors appreciate the data on ternary compound formation provided by Z. Tomczuk. The authors are particularly grateful for the assistance and comments provided by S. P. Perone of Purdue University and S. Wood of Illinois Institute of Technology.

Literature Cited

1. Cairns, E. J., Shimotake, H., *Science* (1969) **164**, 1347.
2. Walsh, W. J., Argonne National Laboratory, private communication, 1973.
3. Sharma, R. A., General Motors Research Laboratories, private communication, 1973.
4. Smirnov, M. V., Podlesnyak, N. P., *Zh. Prikl. Khim.* (1970) **43**, 1463.
5. Watanabe, N., Nakanishi, K., Komura, A., Nakajima, T., *Kogyo Kagaku Zasshi* (1968) **71**, 1599.
6. Heus, R. J., Egan, J. J., *J. Phys. Chem.* (1973) **77**, 1989.

7. Egan, J. J., Brookhaven National Laboratory, private communication, 1973.
8. Snyder, R. N., Lander, J. J., *Electrochem. Technol.* (1966) **4,** 179.
9. Cairns, E. J., General Motors Research Laboratories, private communication, 1973.
10. Jordan, D. O., Lane, J. E., "The Wetting of Solid Metals by Liquid Alkali Metals," in "The Alkali Metals," p. 147, The Chemical Society, London, 1967.
11. Bodewig, F. G. Plambeck, J. A., *J. Electrochem. Soc.* (1969) **116,** 607.
12. Bernard, J. P., DeHaan, A., Van der Poorten, H., *C. R. Acad. Sci., Ser. C* (1973) **276,** 587.
13. Bernard, J. P., Electrochemistry Laboratory, Mons, Belgium, private communication, July 5, 1973.
14. Kennedy, J. H., Adamo, F., *J. Electrochem. Soc.* (1972) **119,** 1518.
15. Cleaver, B., Davies, A. J., Schiffrin, D. J., *Electrochim. Acta* (1973) **18,** 747.
16. Liu, C. H., Zielen, A. J., Gruen, D. M., *J. Electrochem. Soc.* (1973) **120,** 67.
17. Merritt, M. V., Sawyer, D T., *Inorg. Chem.* (1970) **9,** 211.
18. Gruen, D. M., McBeth, R. L., Zielen, A. J., *J. Amer. Chem. Soc.* (1971) **93,** 6691.
19. Vogel, R. C., et al., "Chemical Engineering Division Research Highlights, January-December 1971," (1972) **ANL-7850,** 146.
20. Giggenbach, W., *Inorg. Chem.* (1971) **10,** 1306.
21. *Ibid.*, 1308.
22. Cleaver, B., Davies, A. J., *Electrochim. Acta* (1973) **18,** 741.
23. Oei, D.-G., *Inorg. Chem.* (1973) **12,** 438.
24. Cunningham, P. T., Johnson, S. A., Cairns, E. J., *J. Electrochem. Soc.* (1972) **119,** 1448.
25. Sharma, R. A., *J. Electrochem. Soc.* (1972) **119,** 1439.
26. Vogel, R. C., et al., "Chemical Engineering Division Annual Report 1970," (1971) **ANL-7775,** 120.
27. Heredy, L. A., Atomics International Division of Rockwell International, private communication, 1973.
28. Berkowitz, J., "Molecular Composition of Sulfur Vapor," in "Elemental Sulfur, Chemistry and Physics," (B. Meyer, Ed.), Chap. 7, p. 149, Interscience, New York, 1965.
29. Hansen, M., "Constitution of Binary Alloys," pp. 176-177, McGraw-Hill, New York, 1958.
30. Chizhikov, D. M., Shchastlivy, V. P., "Selenium and Selenides," trans. by E. M. Elkin, Chap. 12, pp. 355-361, Collets, London, 1968.

RECEIVED May 1, 1974. This work was performed under the auspices of the U.S. Atomic Energy Commission.

13

Metal Sulfide Electrodes for Secondary Lithium Batteries

LASZLO A. HERÉDY, SAN-CHENG LAI, LOWELL R. McCOY, and RICHARD C. SAUNDERS

Atomics International Division, Rockwell International Corp., Canoga Park, Calif. 91304

> *Electrochemical properties of several transition metal sulfides were studied in cells using lithium anodes and LiCl–KCl eutectic salt electrolyte at 400°C. Copper and iron sulfides, studied in detail, are reversible and high use of theoretical capacity is obtained. Two 5-mo cycle tests made with lithium–copper sulfide cells showed good capacity retention. Coulombic efficiencies, initially over 95%, suffer a small loss with time because of lithium dispersions formed at the anode surface, causing self-discharge of sulfide cathodes. Coulombic efficiency can be restored by replacing the lithium anode and electrolyte. Design projections indicate that energy densities of 77 W-hr/lb can be attained with lithium–iron sulfide batteries and 52 W-hr/lb with lithium–copper sulfide batteries.*

Bulk energy storage for electric utility load-leveling and all-electric vehicles offers very large potential markets for new low-cost, high energy density batteries (*1, 2*). These requirements are most likely to be met by high temperature molten salt or solid electrolyte batteries now under development at a number of laboratories. One of these, the lithium–sulfur battery, has been actively investigated for several years (*3, 4, 5*). A general discussion of lithium–sulfur cells as well as other high temperature cells was published by Cairns and Steunenberg (*6*). Also, several summaries of experimental results obtained with lithium–sulfur cells and battery design estimates are available (*3, 4, 5*).

It was found in these studies that the lithium–sulfur cell system could meet many of the requirements needed to develop a high energy density battery system. Long-term cycling tests showed, however, that the operating life of the sulfur electrode was not satisfactory because of serious basic difficulties related to containing sulfur electrode discharge products. In particular, it has not been possible to retain over extended periods the lithium polysulfides, which form during discharge in the cathode compartment. Since they are soluble (7) in the electrolyte, they diffused through porous separators and reached the lithium electrode where they were converted to the sparsely soluble Li_2S. Under these conditions, the cathode gradually lost capacity and could not be fully recharged under practical operating conditions.

This paper presents the results of recent investigations on an alternative class of cathode materials, the transition metal sulfides. These exhibit good reversible electrochemical behavior and, unlike elemental sulfur cathodes, retain their capacity after prolonged charge–discharge cycling.

Exploratory Tests

During early work with elemental sulfur electrodes, several additives were used to improve the retention of sulfur or lithium polysulfides by adsorption or the formation of weak chemical bonds. None of the additives worked satisfactorily. While the rate of capacity loss was decreased in several instances, it still remained unacceptably high. It was noted, however, that a small residual cathode capacity remained when molybdenum or molybdenum sulfide was used as the additive. The latter effect was attributed to the electrochemical activity of molybdenum–sulfur compounds formed during cycling of the sulfur electrode. An experiment was therefore made using the compound K_2MoS_4 as the active cathode material. This electrode sustained over 300 charge–discharge cycles at 50 ma/cm^2 current density with no capacity loss.

The favorable results obtained with K_2MoS_4 suggested that other metal sulfide compounds could also exhibit good electrochemical activity and reversible behavior in molten LiCl–KCl electrolyte at 400°C in contrast to the known low electrochemical activity and poor rechargeability of metal sulfides in organic electrolytes at room temperature. A review of thermodynamic data, physical properties, and calculated theoretical energy densities of sulfur compounds identified the transition metal sulfides as the most promising group of cathode materials. It was calculated that, depending on the composition of the particular metal sulfide, the open circuit voltages of lithium–metal sulfide cells would vary from approximately 1.3 to 2.0 V and theoretical energy densities would be in the range of 250 to 600 W-hr/lb.

The cell reaction for a bivalent metal sulfide and metallic lithium couple can be written as follows:

$$2\ Li + MeS = Me + Li_2S$$

The electrochemical cell can be shown as:

$$- \text{Li (l)} \mid \text{LiCl-KCl} \mid \text{MeS (s)} +$$

Exploratory tests with iron, copper, nickel, and manganese sulfides were performed in molten LiCl–KCl eutectic mixture at 420°C, using simple electrode structures containing metal sulfide powders or metal powder and lithium sulfide (all <320 mesh). These were mixed with fine conductive carbon or graphite powder. The sulfide electrode, together with a lithium counter-electrode, was immersed in the electrolyte, and the cell was cycled at different current densities. Table I shows the reversible voltage plateaus and corresponding coulombic efficiencies observed during the cycling tests. The best coulombic efficiency was obtained with copper sulfide, but quite good efficiencies were obtained with nickel and iron sulfides also. The electrodes worked equally well when the starting material was a metal sulfide (CuS or FeS_2) or a mixture of metal and lithium sulfide ($Ni + Li_2S$). It was concluded from the exploratory measurements that further investigation of iron, copper, and nickel sulfides was definitely warranted. On the other hand, studies on manganese sulfide were postponed because of the low coulombic efficiency obtained with that material. The reason for the low coulombic efficiencies observed in some instances was not understood at that time. One explanation is described below (Figure 3).

Table I. Exploratory Test Results of Lithium–Metal Sulfide Cells

Composition of Cathode Mixture	Reversible Voltage Plateau[*] (v)	Coulombic Efficiency (%)
FeS_2 + Carbon Powder[†]	2.03	60
	1.62	80
CuS + Carbon Powder	1.68	95
Ni + Li_2S + Carbon Powder[†]	1.85	70
	1.70	90
Mn + Li_2S + Carbon Powder	1.55	25

[*] Open circuit cell voltage
[†] Two voltage plateaus observed

Detailed Electrode Tests

Extensive testing of metal sulfide electrode materials was performed in cells of the type shown in Figure 1. The metal sulfide electrode case and current take-off rod were made of dense graphite to avoid any contact and possible contamination of the active material by metallic constituents. The active material (metal sulfide powder in a porous graphite matrix or a mixture of metal sulfide and graphite powder) was contained in the cavity (2.5 cm in diameter $\times$ 0.6 cm deep) of the dense graphite electrode case. A porous ceramic separator was placed over the active material and secured with high purity alumina pins.

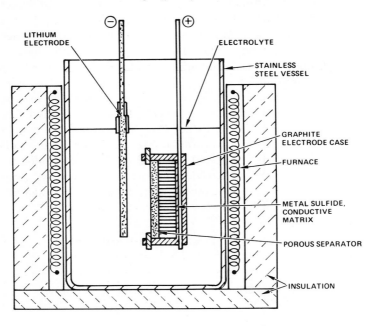

Figure 1. Lithium–metal sulfide investigative cell

Lithium electrodes were made of square stainless steel, 90% porous plates, ~10 cm² in area by 0.3 cm thick. These were impregnated with lithium at 650°C prior to use.

The lithium and metal sulfide electrodes were immersed in ~250 ml of an electrolyte consisting of 58.8 mole % LiCl and 41.2 mole % KCl, saturated with Li_2S and maintained at 380–400°C by an electric furnace. Typical examples of cycling curves obtained with different metal sulfide electrodes are shown in Figures 2 and 3. The tests shown in Figure 2 and the upper curve in Figure 3 were made with sulfide electrodes with graphite cases. The lower cycle curve in Figure 3 was recorded with an

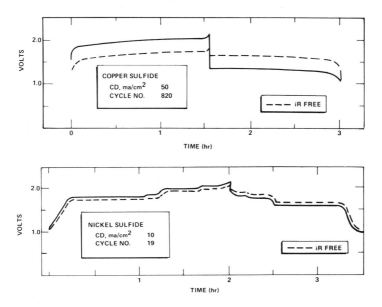

Figure 2. Charge–discharge records for lithium–copper sulfide and lithium–nickel sulfide cells

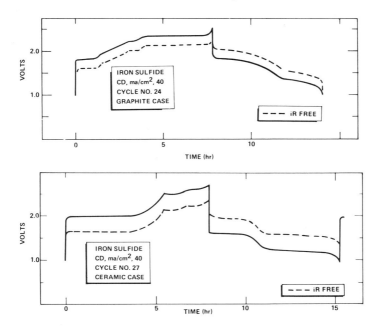

Figure 3. Charge–discharge records for lithium–iron sulfide cells with graphite and ceramic cases

FeS$_2$ electrode enclosed in a porous ceramic case. The significance of this latter test will be discussed later in a separate section.

Table II summarizes the results of cycle tests made with metal sulfide electrodes using dense graphite cases. In general, 50–200 cycles were completed in one test. The cell voltages and the coulombic efficiencies reached the values shown after the first few cycles and did not change further during the tests. It has been established by x-ray diffraction analysis of the discharged electrodes that each of the four metal sulfides was reduced to the corresponding metal during discharge. The active material compositions in the charged state were calculated from the coulombic efficiencies and the currents passed during the charging period.

Table II. Properties of Metal Sulfide Active Materials

Property	Iron Sulfide		Chromium Sulfide	Nickel Sulfide	Copper Sulfide
Open circuit voltage (V)[a]	2.03	1.62	1.27	1.70[b]	1.68
Approximate composition of active material:					
Charged state	FeS$_{1.5}$	FeS	CrS	Ni$_3$S$_2$	Cu$_2$S
Discharged state	Fe	Fe	Cr	Ni	Cu
Coulombic efficiency (%)[c]	75	85	90	85	95
Theoretical energy density of lithium–metal sulfide cell (W-hr/lb)	525[d]	386	316	308	236

[a] Voltage vs lithium electrode.
[b] Main voltage plateau; for other plateaus see Figure 2.
[c] Determined at C/4 cycling rate.
[d] Based on average discharge voltage of 1.8 V.

In the case of iron and nickel sulfides, the coulombic efficiencies can be quite different at the successive voltage plateaus as the highest valence metal sulfide is reduced stepwise to the metal (see Table I). In general, higher coulombic efficiencies were obtained if charging was stopped at a lower valence state. For example, cycling between Fe and FeS yielded coulombic efficiencies averaging ~80% as compared with the 60% obtained in cycling between FeS and FeS$_{1.5}$ under otherwise similar conditions.

On the basis of the results summarized in Table II, copper and iron sulfides were selected for more detailed testing. Copper sulfide (Cu$_2$S) showed good electrochemical activity and reversible behavior and the best coulombic efficiency of the metal sulfides investigated. Iron sulfide (starting with FeS$_2$) also exhibited good electrochemical activity and reversible behavior, but the coulombic efficiency obtained with this material was less than that found for copper sulfide. The theoretical energy density of the lithium–iron sulfide couple is significantly higher than that of the lithium–copper sulfide couple. Because of its high coulombic efficiency, copper sulfide was selected for the active material for initial testing, for long-term cycling tests, and for electrode scale-up studies while the iron sulfides were explored further.

Copper Sulfide Electrode Tests

Several cycling tests were made with copper sulfide electrodes in investigative cells of the type shown in Figure 1. The active cathode material, in most cases Cu_2S, was distributed in a 90% porous graphite matrix held in the dense graphite housing. In experiments where CuS was used as the starting active material, x-ray diffraction measurements of active material taken from cycled electrodes showed that it was converted almost entirely to Cu_2S during the first discharge–charge cycle. The use of active material and coulombic efficiency were determined in cycling experiments of 50–200 cycles for electrodes having different specific storage capacities. The specific storage capacity, amp-hr/cm^3, was calculated from the theoretical capacity of the weight of active material present in the electrode and the volume of the electrode cavity in which the active material and porous graphite matrix were held. In the specific storage capacity range of 0.20–0.38 amp-hr/cm^3 investigated, active material utilizations of 88–95% and coulombic efficiencies of 85–98% were obtained. The active material in fully charged electrodes was examined after two experiments by x-ray diffraction. It was found in both cases that the x-ray pattern of the main component was that of Cu_2S. In addition, the active material also contained sizeable amounts of $4Cu_2S \cdot CuS$.

Two long-term tests have been performed with copper sulfide electrodes, one in a small investigative cell of the type shown in Figure 1, the other in a sealed 25-W-hr cell which will be described later. The starting active material in both cells was CuS. In both cases, the lithium electrodes consisted of porous metal plates impregnated with molten lithium metal. The electrolyte was the LiCl–KCl eutectic salt mixture, saturated with Li_2S and maintained at 380–400°C.

The history of the investigative cell test is shown in Table III. Most of the apparent losses in the capacity of the electrode and the coulombic efficiency of the cell observed after the 500th cycle were restored by replacing the lithium electrode and electrolyte. Since the apparent capacity loss was only temporary, it was concluded that it could not be attributed to changes in the structure of the active material or the electrode case. The most probable cause of reduced storage capacity and coulombic efficiency was the generation of finely dispersed lithium at the anode and the oxidation of these lithium droplets on the cathode surface. Insertion of a fresh lithium electrode and replacement of the electrolyte eliminated lithium dispersion from the cell, and thus the original cell performance was restored. The lithium electrode was not capacity-limiting prior to its replacement. The test had to be terminated after the 912th cycle because of a glovebox accident which seriously contaminated the atmosphere with air and water vapor. The electrolyte bath was also considerably contaminated with water, and fine cracks developed in the electrode case.

Table III. Copper Sulfide Electrode Cycling Test Histories

Event	Cycle Number	Capacity (amp-hr)	Coulombic Efficiency
0.6-whr Investigative Cell			
Startup/Equilibrium	15	0.385	97
Status, 72nd Day	400	0.38	85
Status, 92-95th Day	500-520	0.34*	80*
New Lithium Electrode	540-560	0.36*	85*
New Electrolyte	600	0.38	97
Status, 136th Day	650	0.38	96
Status, 153rd Day	750	0.37	94
Status, 203rd Day	912	0.36	88
25-whr Demonstration Cell			
Startup/Equilibrium	5	17.2	97
Status, 47th Day	110	16.4	97
Status, 74th Day	170	15.4	95
Status, 97th Day	225	14.2	85
New Lithium Electrode	238	16.0	90
Experiment Terminated, 108th Day	244	16.0	90

*Average values

The history of the 25-W-hr cell, also shown in Table III, is similar to that of the investigative cell. Diminished capacity after 225 cycles was restored by lithium electrode replacement. As in the previous test, the lithium electrode was not capacity-limiting prior to its replacement. Unlike the investigative cell, however, electrolyte replacement was not attempted, and it is not known whether the coulombic efficiency could have been improved by this procedure.

Iron Sulfide Electrode Tests

Because of the higher energy density of iron sulfide as compared with those of the other metal sulfides shown in Table II, studies have been made to improve the low coulombic efficiency observed in the early investigations. It was theorized that the low coulombic efficiency could be caused by a self-discharge reaction, where the lithium electrode could be the source of finely dispersed particles of lithium which, transported freely by electrolyte convection, could be oxidized on the electronically conductive surface of the graphite case of the metal sulfide electrode. The resulting self-discharge reaction for the reduction of FeS_2 (probably present as $FeS_2 \cdot FeS$) after repeated cycling to FeS is shown in Figure 4. Further reduction of FeS to iron may take place by the same mechanism.

Experiments have been made to test the proposed self-discharge mechanism. Cycling tests have been carried out with iron sulfide electrodes where the active material matrix, finely ground FeS_2 of high purity in porous graphite, and the dense graphite current take-off lead were held within a porous ceramic crucible. On the basis of the proposed mechanism, a significant reduction in self-discharge could be expected from this electrode design, as the use of the porous ceramic barrier should greatly impede lithium transport to the metal sulfide electrode. The experimental results supported the proposed self-discharge mechanism described above. A typical cycling curve obtained with the ceramic-enclosed iron-sulfide electrode is shown as the lower curve in Figure 3. Coulombic efficiencies of 92–96% were obtained in contrast to 60–75% obtained in earlier cycling tests in which the graphite case

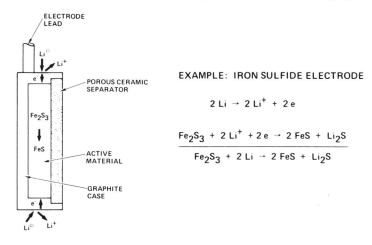

Figure 4. Proposed self-discharge mechanism for metal sulfide electrodes

of the iron sulfide electrode was directly exposed to the lithium dispersion and where rapid self-discharge could take place. Typical cycling tests were of 3- to 6-wk duration (25–50 cycles). Active material utilizations of 90–100% based on $FeS_{1.5}$ were obtained. No performance degradation was observed during these experiments. Long-term testing is being initiated.

Demonstration Cell Test

To obtain information regarding the scale-up of electrodes and the design and construction of large cells, a sealed 25-W-hr lithium–copper sulfide cell was built and tested. The schematic of the cell is shown in Figure 5. A photograph of the cell prior to installation of the insula-

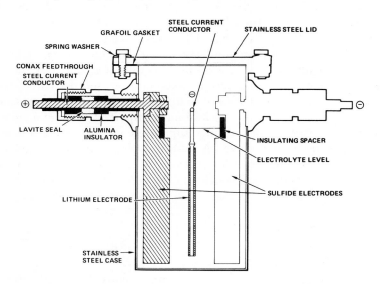

Figure 5. *25-W-hr lithium–metal sulfide demonstration cell*

tion appears in Figure 6. The history of this test is shown in Table III. Cycling was generally performed at a 4-hr rate (at a current density of ~30 ma/cm²) in both the charge and discharge modes. Between the 80th and 145th cycles, the current density was varied in four steps from 10–37 ma/cm² to investigate the correlation between current density, cell capacity, and coulombic efficiency. The results are shown in Table IV. After an operational period of 3½ months, the test was terminated. Inspection of the cell showed no marked corrosion of the cell case or deterioration of the lithium electrode. A detailed study of the sulfide electrodes showed no attack either on the dense graphite structures or on the porous

Figure 6. 25-W-hr lithium–metal sulfide cell prior to installing insulation

graphite current collectors. Some attack was evident on the electrolyte (lithium) side of the porous alumina separator. The appearance of the copper sulfide active material and its distribution in the electrode cavity were not significantly altered from the original conditions.

Projected Battery Characteristics

Energy density projections have been made for lithium–metal sulfide battery modules sized for use in load-leveling energy storage plants on the basis of the experimental results obtained to date. The results of calculations made for modules using Cu_2S, FeS, and FeS_2 as the active cathode materials are shown in Table V. Volume capacity design goals,

Table IV. Test Data for a 25-W-hr Demonstration Cell

Current Density (ma/cm^2)	Discharge Rate (hr)	Capacity (ah)	Coulombic Efficiency (%)
10	12	17.6	93
18	7	17.1	96
30	4	16.4	97
37	3	15.3	98

expressed in amp-hr/cm³, and those achieved experimentally with two metal sulfides are also shown. Depending on the metal sulfide selected, energy densities up to 77 W-hr/lb and 9 W-hr/in.³ are believed attainable. Energy (watt-hr) efficiencies in the range of 75% are expected for batteries designed for load-leveling applications.

Table V. Battery Module Characteristics

Parameter/Metal Sulfide	Cu_2S	FeS	FeS_2
Open Circuit Voltage (v)	1.68	1.62	2.03
Operating Voltage (v)	1.4	1.4	1.55
Cathode Capacity (amp-hr/cm³)*	0.80	1.0	1.10
Energy Densities			
(whr/lb)	52	66	77
(whr/in.³)	6.7	7.8	9.2

*Capacities Experimentally Achieved: 0.35 for Cu_2S, 0.55 for FeS_2

Conclusions

Transition metal sulfides exhibit good reversible electrochemical behavior in molten KCl–LiCl eutectic salt electrolyte at 400°C. Copper-sulfide cathodes in lithium/Cu_2S cells retain their capacity after prolonged charge–discharge cycling. Use of over 90% of theoretical capacity and coulombic efficiencies in excess of 95% are obtained initially. Modest losses of cathode capacity and coulombic efficiency after hundreds of cycles result from gradual deterioration in the performance of the lithium electrode. The generation of fine lithium dispersions at the anode causes accelerated self-discharge of the cathode. Both cell capacity and coulombic efficiency can be restored by replacing the lithium electrode and electrolyte. The excessive rate of self-discharge of iron sulfide electrodes observed in early investigations is also traceable to dispersed lithium formation at the anode.

The transition metal sulfides, notably iron and copper sulfides, are relatively inexpensive and abundant and are, therefore, well suited for use in large-scale battery applications such as bulk energy storage for electric utility load-leveling facilities. Energy densities approaching 80 W-hr/lb and 9 W-hr/in.³ are projected for lithium–metal sulfide batteries designed for this application.

Acknowledgment

The writers wish to acknowledge the assistance of Michael Klenck, Atomics International, Chemistry Technology Department, who made the x-ray diffraction analyses of copper sulfide materials.

Literature Cited

1. Rosengarten, W. E., Jr., et al., "Wanted: Load-Leveling Storage Batteries," *Extended Abstracts of Electrochem. Soc.* (1972) **72-2**, paper No. 34.
2. George, J. H. B., et al., "Prospects for Electric Vehicles. A Study of Low Pollution Potential Vehicles—Electric," Dept. of Health, Education and Welfare, Arthur D. Little, Cambridge, Mass., May 15, 1968.
3. Kyle, M. L., et al., "Lithium/Sulfur Batteries for Off-Peak Energy Storage," Argonne National Laboratory, Report **ANL-7958**, March 1973.
4. Heredy, L. A., Parkins, W. E., "Lithium–Sulfur Battery Plant for Power Peaking," Paper No. **C-72-234-8**, IEEE Meeting, New York, February 4, 1972.
5. Kyle, M. L., et al., "Lithium–Sulfur Batteries for Electric Vehicle Propulsion," Paper No. 38, Intersociety Energy Conversion Engineering Conference Proceedings, 1971.
6. Cairns, E. J., Steunenberg, R. K., "High Temperature Batteries," in "Progress in High Temperature Physics and Chemistry," Vol. 5, C. A. Rouse, Ed., pp. 63–123, Pergamon, New York, 1973.
7. Kennedy, J. H., Adams, F., *J. Electrochem. Soc.* (1972) **119**, 1518.

RECEIVED July 17, 1974. This work was funded jointly by the Edison Electric Institute (now the Electric Power Research Institute), the Tennessee Valley Authority, and Rockwell International Corp., Atomics International Division.

14

Sodium–Sulfur Batteries

LYNN S. MARCOUX and EDDIE T. SEO

TRW Systems Group, TRW Inc., One Space Park, Redondo Beach, Calif. 90278

> *The sodium–sulfur battery is one of the high energy electrochemical devices being developed to store energy. These batteries satisfy the specific energy requirements for both automotive and utility load-leveling applications. Sodium–sulfur cells which use β-alumina cell dividers have unique properties other than high specific energy, and work is being carried out by many organizations to develop practical battery systems.*

A growing awareness of the energy requirements of a technologically advanced society coupled to a deepening understanding of the environmental implications of that technology has renewed interest in electrochemical energy storage devices. This interest has manifested itself in a growing effort to develop high energy battery systems for energy storage. These devices will probably be used mainly as batteries for electric vehicles and as load-leveling storage systems for electric utilities.

The specific requirements of these two applications differ in several respects, but generally both require long-lifetime, high specific energy batteries which operate in a nonpolluting manner. Specific energy is a prime requirement for vehicular traction applications whether in small urban vehicles or electric railways. A workable goal for this application is a battery which provides a specific energy of 150–200 W-hr/kg. Load leveling—the storage of electrical energy generated during off-peak hours to be used during peak consumption periods—does not impose as stringent a weight requirement. Nonetheless the battery installation must be as compact as possible so that the batteries may be sited at the sub-station, thereby greatly reducing transmission costs. Several electrochemical systems are being developed to meet this requirement.

The selection of systems which might be suitable for high specific energy batteries begins with the consideration of chemical periodicity. These purely thermodynamic considerations must of course also be tempered by the limitations of kinetics as well as by the various requirements of working batteries such as cost and fabricability. These analyses indicate that sulfur, although not as energetically endowed as some of its neighbors on the periodic chart, is a sufficiently powerful oxidant for secondary battery applications. Low cost, ready availability, and an extensive technology for handling also make this element attractive.

The alkali metals—lithium, sodium, and potassium—are logical choices for anodes in a sulfur-based electrochemical cell. All three have been incorporated into cells, and lithium and sodium remain under serious consideration. The lithium–sulfur combination is the topic of another chapter in this volume and will not be discussed further. Two types of sodium–sulfur cells have been constructed. One type uses thin-walled glass capillaries as a cell divider, and the other uses various sorts of ionically conducting sodium aluminate for this purpose. Of the two, the latter seems to hold the most promise and certainly has generated the most interest and enthusiasm (1). Because of the unique properties of the solid electrolyte cell separator this battery is also probably the most interesting from a purely scientific point of view.

This paper restricts itself to sodium–sulfur cells and batteries which use solid electrolyte cell dividers and provides a current picture of the state of scientific knowledge and technological achievement with respect to sodium–solid electrolyte–sulfur batteries. The references cited should not be construed as a complete review but should instead be viewed as an introduction to the relevant literature.

Basic Properties of Sodium–Solid Electrolyte–Sulfur Batteries

The sodium–sulfur cell was first demonstrated by Kummer and Weber at the scientific laboratory of the Ford Motor Co. (2). The first cell, as its creators point out, was not the outgrowth of a conscious search for a battery system but was instead a logical development in the pursuit of ionically conducting glasses. The study of various glasses led to a ceramic material with a composition best expressed by $Na_2O \cdot 11\ Al_2O_3$ and known as β-alumina. This material was well-known, but its high ionic conductivity had not been recognized. At 300°C the sodium ion conductivity is of the same order of magnitude as molten salts. This original work led to several patents (3).

It is the unusual properties of the solid electrolyte which permit the coupling of sodium and sulfur in an electrochemical device. The sodium ion mobility of this material results from a crystal structure in which layers of mobile sodium ions alternate with layers of close-packed blocks of aluminum and oxygen ions. These layers occur in slightly different ratios and symmetries resulting in more than one crystalline modification.

Table I. Objectives for Sodium Aluminate Solid Electrolytes

Property	Desired Value
Fracture Stress	3–4 × 10^4 psi
Density	>98% of theory
Resistivity	10 Ω cm at 300°C
Cost	$3.30/ft^2

The best known crystalline forms are designated β- and β''-alumina. The structural variations added to the fact that these materials are non-stoichiometric give the name β-alumina almost a generic meaning. The chemistry and physics of these materials have come under close scrutiny, and a large literature has taken shape (4).

The development of β-alumina for use in long-life batteries consists of more than simply maximizing the sodium ion conductivity. It has, in fact, been pointed out (5) that the conductivity specification is not an independent variable since by virtue of the conductivity mechanism, strength and conductivity are related. Greater strength occurs with smaller grain size; however, resistance arising at the grain boundaries increases. For example the resistivity of a single crystal of β-alumina is 4.3 Ω cm at 300°C (6) while for a comparable polycrystalline sample it is greater than 15 Ω cm at the same temperature. The goal of those involved in solid electrolyte development is to develop a reproducible material with both the requisite structural and electrical properties. For electric utility load-leveling applications, Birk, of the Electric Power Research Institute (EPRI), has quantified these objectives as shown in Table I (5).

As well as having strength and conductivity, the electrolyte must also endure both thermal and electrical cycling for extended time periods (10–20 yrs) in order to provide satisfactory service. For this reason electrolyte failure modes have been a continuing research topic. Several possible electrolyte failure mechanisms have been advanced. One suggested mechanism is that electrolytically deposited sodium exerts pressure in electrolyte pores (7, 8). This mechanism is supported by the results of Auger electron spectroscopy which indicated excess sodium present at the fracture surface (9, 10). The incorporation of impurity ions into the β-alumina lattice is another possible failure mechanism. Cells assembled with high purity solid electrolyte have sustained over 3000 cycles and 1450 A-hr/cm^2 of charge (11, 12). The study of the mechanism and prevention of electrolyte failure will continue to have high priority.

A diagram of a recent sodium–sulfur cell is shown in Figure 1 along with a photograph of the actual cell. This cell was constructed from interchangeable hardware and was designed for laboratory studies rather than as a component cell in a practical battery. The cell consists of a central tube of β-alumina which contains sodium. This tube is immersed in a graphite felt impregnated with sulfur. The graphite felt serves as the actual cathode in the system. At the operating temperature, 300°C, both the sodium and the sulfur are molten. The cathode current collector is unnecessary in cells which use a stainless steel cathode container. Other cells based on a tubular design are similar to this, differing only in the details of assembly. Most published reports concern tubular cells. However, cells based on flat plates of solid electrolytes have been built and possess many advantages for certain applications (*13, 14, 15*).

Cell operation most simply stated consists of sulfur reduction by sodium, with sodium ions being transported through the solid electrolyte

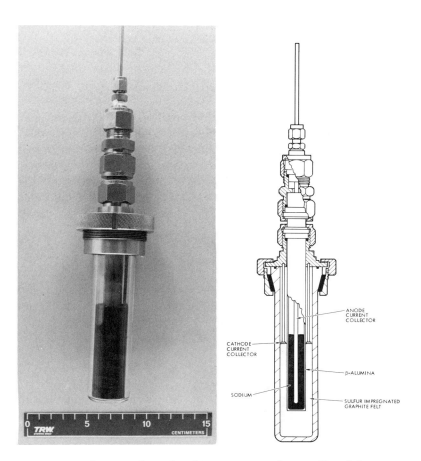

Figure 1. Photograph and schematic of sodium–sulfur laboratory test cell

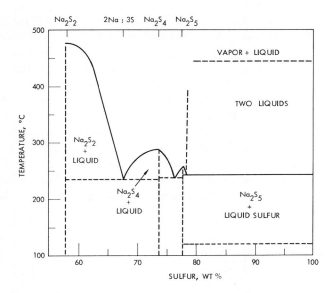

Figure 2. Phase diagram of the sodium–sulfur system

to form sodium polysulfides while electrons flow in the external circuit. The reverse process occurs with 100% coulombic efficiency, and this constitutes one of the major advantages of the sodium–sulfur system. Since no metal plating reactions are involved in the charging sequence, dendrite growth and plate reformation problems which often plague secondary batteries are not a concern. Degradation of the solid electrolyte does, however, occur with prolonged use. Although the discharge reaction is frequently represented by:

$$2\text{Na} + 3\text{S} \rightarrow \text{Na}_2\text{S}_3$$

this is an oversimplification and more correctly represents the stoichiometry of discharge for a practical cell. Because of the existence of several polysulfide species over the discharge range the reaction is better expressed as:

$$2\text{Na} + x\text{S} \rightarrow \text{Na}_2\text{S}_x$$

The physical, chemical, and electrochemical properties of the sulfur–sodium polysulfide system influence many aspects of cell performance. For example, the 300°C operating temperature is dictated by the requirement that reaction products be liquid, and it is likewise this requirement which ultimately limits the extent of cell discharge. The phase behavior of sulfur–sodium polysulfide melts has been determined by both differential thermal analysis (16) and from open-circuit cell voltages (17). From these data an accurate phase diagram, Figure 2, has been constructed. The salient features of this diagram are:

1. The battery must be operated at temperatures near 300°C to avoid safely the precipitation of sodium polysulfides.
2. The much earlier phase diagram of Pearson and Robinson (18) is partially in error.
3. The species Na_2S_3 is not stable above 100°C, and higher temperature mixtures of this composition consist of a 1:1 eutectic of Na_2S_4 and Na_2S_2. A related study (19) has also demonstrated the nonexistence of Na_2S_6.

Phase behavior is partially reflected in the battery discharge curves shown in Figure 3. The battery was a 6-V stack consisting of three plate-type cells which used 2.5-cm diameter solid electrolyte disks (13). The polarization is almost entirely caused by the combined ohmic resistance of the solid electrolyte and the sulfur–sodium polysulfide melt. The high polarization near the point of complete discharge is caused by the precipitation of Na_2S_2. Complete discharge curves exhibit a high initial polarization resulting from the high resistivity of pure sulfur (20). The resistivities of sulfur (21) and molten polysulfides (22) have been measured and tabulated over a wide temperature range and confirm the above interpretation. An interesting aside to these results was the devel-

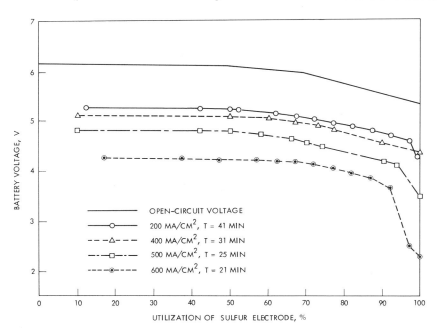

Figure 3. Discharge curves of a 6-V sodium–sulfur battery

opment of a technique which permitted the preparation of very pure solutions of sodium polysulfide in sulfur *via* electrolysis.

Other physical properties of the sulfur–polysulfide melt influence cell performance and have been studied. Density is perhaps foremost among these since density variation during discharge will dictate cathode void volume requirements. Furthermore since the sulfur–polysulfide melt is a two-phase system (*see* Figure 2) over much of the discharge range, density in conjunction with viscosity and surface tension establish the nature of the phase separation. Measurements of density (*19, 23*) demonstrate that the reaction proceeds at 360°C according to:

$$2Na(l) + 5S(l) \rightarrow Na_2S_5(l)$$

53.5 cm³   96.4 cm³   111 cm³

indicating a decrease in total volume but an 11.5% increase in catholyte volume. Regression equations were developed which permit the calculation of density as a function of composition and temperature. At 360°C the densities of sulfur and sodium pentasulfide are 1.66 g/cm³ and 1.86 g/cm³, respectively. These results are consistent with earlier electrochemical measurements (*24*) which indicated two liquid phases the lower of which was polysulfide-rich. This phase separation must be taken into account since it is likely that the point of electrochemical reaction is at the carbon–sulfur–sodium polysulfide three-phase junction. The obvious similarity to fuel cells has been noted (*26*), and cathode management is a growing concern among cell designers. One approach to the problem is to operate the cell only in the single-phase region (*25*). Unfortunately, this approach dramatically decreases the specific energy of the system.

The sulfur–sodium polysulfide system has received the attention of electrochemists but few of the studies have been under conditions comparable to sodium–sulfur battery operating conditions. The thermodynamics of the system have been studied by means of open-circuit potentials (*17, 27*), and dynamic measurements have been made in fused salts (*28*). The most pertinent studies are those of sulfur–polysulfide electrochemistry in the actual sulfur–polysulfide melts (*24, 29, 35*). The results of these studies seem to indicate that both the oxidation and reduction reactions are rapid, although the oxidation reaction is hindered by the formation of an insulating sulfur film. These studies also concluded that the electrode reaction sequences were quite complex because of the multitude of polysufide species. As the system becomes better characterized more quantitative descriptions are possible as evidenced by a recent work which modeled the resistive drop through an actual sulfur impregnated graphite electrode in order to correlate the spatial distribution of reaction products (*30*). Clearly more complete descriptions of this type for the sulfur cathode are necessary to optimize cell and battery design.

The construction of operational, hermetically sealed sodium–sulfur cells requires container materials, which are mechanically suitable and compatible with sodium and sulfur–sodium polysulfide, current leads to the sulfur–graphite electrode, and several kinds of seals. These requirements of course are in addition to those for the solid electrolytes and sulfur electrodes described earlier. The problem of satisfying these requirements has been summarized by Gratch and co-workers (*3*). Particular applications—utility load-leveling, traction power, and military—impose further design constraints dictated by performance, capacity, size and weight, life, cost, and safety requirements.

Battery Development

With respect to the present state of battery development, it can be simply stated that most of the current work is being carried out at the laboratory level and that it is concentrated towards acquiring the necessary technology for the design and construction of practical cells and battery systems. However, there have been several successful attempts to build hermetically sealed cells and to assemble a large number of these cells into batteries. Four notable examples are discussed below.

Ford Motor Co. has built and tested a 24-cell battery containing four parallel strings of six tubular electrolyte cells in series (*3*). The 12-V, 400-W (peak power) battery's power and energy exceeded 209 W/kg

Figure 4. Sodium–sulfur battery modules

and 95 W-hr/kg, respectively, exclusive of thermal insulation and packaging. Results indicate that in the Ford battery design, the specific power and specific energy (power and energy densities) must be optimized separately. Ford cells have achieved longevity in excess of 2000 deep charge–discharge cycles (3).

The Electricity Council Research Centre has developed a 960-cell (40 modules of 24 tubular electrolyte cells each), 50 kW-hr, 100-V traction battery. The modules are shown in Figure 4. The battery was rated at 15.5 kW of average power, 29 kW of peak power, and weighed 800 kg. The specific peak power and specific energy are therefore 36 W/kg and 63 W-hr/kg, respectively.

TRW has built a hermetically sealed 20-A hr cell operable at 300°C and designed to deliver 10 A for 2 hrs above 1 V with a specific power and a specific energy of at least 110 W/kg and 110 W-hr/kg, respectively (Figures 5 and 6) (14, 15). Although, for various reasons, the actual fabricated cells failed to meet the designed performance goals, it was demonstrated that the design was valid. These cells also demonstrated the use of hot-pressed β-alumina materials in a flat-plate configuration.

General Electric Co. has been actively testing tubular cells with the following specifications (31):

Total weight:	156 g
Diameter:	2.54 cm
Length:	13.3 cm
Capacity:	17 A hr
Discharge current:	2 A (0.1 A/cm^2)
Charge current:	1 A (0.05 A/cm^2)
Energy density:	0.45 W-hr/cm^3
Specific energy:	45 W-hr/kg

These cells contain solid electrolytes fabricated by sintering green ware prepared by electrophoretic deposition.

Figure 5. Sodium anode assembly. Left; hot-pressed β-alumina cup. Right; complete anode compartment comprising two mating cups and a fill-tube assembly.

Figure 6. 20-A hr sodium–sulfur cell

Although the cited examples are far from being prototypes of production devices, they do validate the feasibility of fully packaged sodium–sulfur cells and provide a measure of projected performance. It seems that sufficient information is now available to permit researchers to speculate on the design of large cells and battery systems and to project the performance of such devices on the basis of anticipated technology.

The material reviewed in this paper reflects the efforts of many industrial and academic organizations with programs for the study or development of sodium–sulfur cells. The significance of the sodium–sulfur system as an advanced battery is indicated by the roster of both performing and funding organizations. A list of these organizations, with an estimate of manpower, has been published and recently updated at Argonne National Laboratory (32, 33). A newcomer to the roster is Chloride Silent Power Limited. It was recently formed by the Electricity Council and Chloride Group Limited. In addition to internal support, many industrial programs in the United States are supplemented with contracts from government agencies and from the electric utilities through its research-supporting organization, *i.e.*, the EPRI.

The Department of Defense, through the United States Army Electronics Command, is supporting a program at TRW (14, 15). EPRI is funding programs directed towards load-leveling battery systems at both General Electric and TRW (31, 34). A major program involving Ford Motor Co., University of Utah, and Rensselaer Polytechnic Institute is being funded for a second year by the National Science Foundation (NSF) (35). NSF, EPRI, and the National Aeronautics and Space Administration also support numerous other study programs on β-alumina type materials.

A frequent criticism of the new high temperature cells has been that the advantages of their very high specific energy were oversold very early in their development. This review should make it clear that high specific energy is but one of the advantages offered by batteries of this type and that they are uniquely qualified to satisfy several future energy storage needs. An important by-product of these systems has been the stimulation of research at the interface of electrochemistry and materials science.

Acknowledgment

The contributions of J. L. Sudworth (British Railways Board), I. Wynn Jones (Electricity Council Research Centre), J. R. Birk (EPRI) S. A. Weiner (Ford Motor Co.), S. P. Mitoff and R. W. Powers (General Electric Co.), and L. J. Rogers (U.S. Army Electronics Command) are gratefully acknowledged.

Literature Cited

1. Sudworth, J. L., *Sulphur Inst. J.* (1972) **8** (4), 12.
2. Kummer, J. T., Weber, N., *Proc. Power Sources Conf.* (1967) **21**, 37.
3. Gratch, S., Petrocelli, J. V., Tischer, R. P., Minck, R. W., Whalen, T. J., *Intersoc. Energy Convers. Eng. Conf., Conf. Proc.*, **7** (1972) 38.
4. Kummer, J. T., *Progr. Solid State Chem.* (1972) **7**, 141.
5. Birk, J. R., Joint Fall Meeting of the American Ceramic Society (Electronics Division) and the National Institute of Ceramic Engineers, *Preprint Paper* **18-EI-74F** (Denver, September 1974).
6. Fielder, W. L., Kautz, H. E., Fordyce, J. S., Singer, J., *NASA Tech. Memo.* **NASA TMX-71546** (1974).
7. Tennenhouse, G. J., Whalen, T. J., "Abstracts of Papers," 75th Annual Meeting of the American Ceramic Society, Cincinnati, April–May, 1973; Paper **15-S4-73**.
8. Armstrong, R. D., Dickinson, T., Turner, T., *Electrochim. Acta* (1974) **19**, 187.
9. Gjostein, N. A., Chavka, N. G., *J. Test. Eval.* (1973) **1**, 183.
10. Stoddart, C. T. H., Hondros, E. D., *Trans. J. Brit. Ceram. Soc.* (1974) **73**, 61.
11. Fally, J., Lasne, C., Lazennec, Y., LeCars, Y., Margotin, P., *J. Electrochem. Soc.* (1973) **170**, 1296.
12. Lazennec, Y., Lasne, C., Margotin, P., Fally, P., 24th Meeting of the I.S.E., *Preprint* (Eindhoven, September 1973).
13. Sudworth, J. L., Hames, M. D., Storey, M. A., Azim, M. F., Tilley, A. R., *Power Sources 3: Res. Develop. Non-Mechan. Elec. Power Sources, Proc. Int. Symp. 8th, 1972* (1973), 1.
14. Seo, E. T., Sayano, R. R., Carroll, D. F., McClanahan, M. L., Silverman, H. P., Final Report, U.S. Army Electronics Command, Contract No. **DAAB-07-72-C-0312**, January, 1974.
15. Seo, E. T., Sayano, R. R., McClanahan, M. L., Silverman, H. P., *Power Sources Symp., Proc.* (1974) **26**, 74.
16. Oei, D. G., *Inorg. Chem.* (1973) **12**, 435.
17. Gupta, N. K., Tischer, R. P., *J. Electrochem. Soc.* (1972) **119**, 1033.
18. Pearson, T. G., Robinson, P. L., *J. Chem. Soc.* (1930) **132**, 1473.
19. Oei, D. G., *Inorg. Chem.* (1973) **12**, 438.
20. Sudworth, J. L., Hames, M. D., *Power Sources 3: Res. Develop. Non-Mechan. Elec. Power Sources, Proc. Int. Symp., 7th 1970* (1971), 227.
21. Steunenberg, R. K., Trapp, C., Vance, R. M., Cairns, E. J., *Advan. Chem. Ser.* (1972) **110**, 190.

22. Cleaver, B., Davies, A. J., Hames, M. D., *Electrochim. Acta* (1973) **18**, 719.
23. Cleaver, B., Davies, A. J., *Electrochim. Acta* (1973) **18**, 727.
24. Selis, S. M., *Electrochim. Acta* (1970) **15**, 1285.
25. Fally, J., Lasne, C., Lazennec, Y., Margotin, P., *J. Electrochem. Soc.* (1973) **120**, 1292.
26. Weber, N., Kummer, J. T., *Intersoc. Energy Convers. Eng. Conf., Conf. Proc.*, **2** (1967), 913.
27. Cleaver, B., Davies, A. J., *Electrochim. Acta* (1973) **18**, 733.
28. Cleaver, B., Davies, A. J., Schiffrin, D. J., *Electrochim. Acta* (1973) **18**, 747.
29. South, K. D., Sudworth, J. L., Gibson, J. G., *J. Electrochem. Soc.* (1972) **117**, 554.
30. Gibson, J. G., *J. Appl. Electrochem.* (1974) **4**, 125.
31. EPRI, Final Report, General Electric Co., **128-1**, Schenectady, New York, July 1974.
32. Kyle, M. L., Cairns, E. J., Webster, D. S., USAEC Report **ANL-7958**, Argonne, Illinois, 1973.
33. Nelson, P. A., Chilenskas, A. A., Steunenberg, R. K., USAEC Report **ANL-8075**, Argonne, Illinois, preprint.
34. EPRI, Final Report, TRW Inc., **127-1**, Redondo Beach, California, October 1974.
35. Final Report, Ford Motor Co., Contract NSF-C805, Dearborn, Michigan, July 1974.

RECEIVED November 4, 1974

INDEX

INDEX

INDEX

A

Additive combinations,
 sulfur— 162, 163, 198, 199
Aggregate(s)
 —asphalt–sulfur mixes, hot 88
 gradation 87, 91
 physical properties of 105
 segregation 90
 suitability evaluation 94
Aging effects on flexural strength . 70
Air
 -atomized sulfur coatings 20
 pollution abatement 51
 voids in mix,
  percent .. 94, 109, 110, 116, 117
 volume, atomizing 45
Alkali metals anodes 217
Alloocimene polysulfides 11
Allotropes, sulfur 3, 21
β-Alumina 218
Aluminate solid electrolytes, sodium 218
Amorphous sulfur 21
Anodic electrochemical reactions .. 196
Anodic scan 194
Anode, alkali metals 217
Anode assembly, sodium 224
Anti-friction compounds 181
Arsenic mixtures, sulfur— 199
Asphalt
 concrete, gravel 115
 content 108, 109, 111
 insulating value of 165
 mixtures, sand-
  sulfur— 86, 97, 103, 105, 121
 mixtures, sulfur–
  (see Sulfur–asphalt)
  physical properties of 105
Atomizing air volume 45

B

Bacterial activity 172
Barnet clay 36
Base layer gravels 146
Base layer sands 145
Bases, pavement structure 97
Battery
 container materials 223
 development 223
 lithium–sulfur secondary 186
 module characteristics 214, 223
 secondary metal 203

Battery (continued)
 sodium–solid electrolyte–
  sulfur 216, 217, 221, 223
 systems, high energy 216
Bentonite 162
Binders, sulfur/asphalt
 (see Sulfur/asphalt binders)
Bleeding surface overlay 98
Block building construction,
 cinder 61
Boiling point elevation 199
Bonding, surface 61
Bresilien stability 141, 144
Brick interface, sulfur– 78
Brick panels, sulfur-bonded
 prefabricated 76
Building, cinder block 61
Building, sulfur 57

C

Canadian sulfur production and
 inventory 155
Capacity loss of sulfur electrode .. 188
Capacity mixes, high bearing 98
Carbon disulfide–insoluble sulfur
 elastomer 21
Carbon working electrode 194
Carsul 37
Castings 99
Cathodic peaks 192
Cell
 high temperature 225
 lithium–metal sulfide .. 206, 207, 212
 lithium–sulfur 186
 sodium–sulfur 219, 225
 test, investigative 209
Ceramic cases for cells 207
Chain extension reaction 178
Charge-discharge records 207
Chromatography, gel permeation .. 16
Chronopotentiometry 189
Cinder block building construction 61
Civil engineering applications of
 sulfur-based materials 154
Clay, Barnet 36
Coatings
 high temperature 26
 polyphenylene sulfide 183
 sulfur (see Sulfur coatings)
Cohesivity of sulfur/asphalt
 binders 136, 138
Cold region testing of sulfur foam
 coatings 167

Compaction 116–118
Compatibility of S/A binders .. 141, 142
Composites, sulfur-containing . 160, 165
Compressive strength 80, 158, 162, 171
Concrete, asphaltic 115
Conditioner 28, 37
Conductivity, thermal 120
Construction materials 75
Construction, road 130
Continuous sulfur wallboard production process 83
Controlled-release properties 33
Copper
  sulfide cells, lithium– 207
  sulfide electrodes 209, 210
  tailings 64
Couple, metallic lithium 204
Creep 160, 161
Cross-linked polymers 2
Crystal structure 217
Crystallinity, sulfur 5, 8, 9
CTLA polymer 65
Curing 109, 178
Cyclic voltammetry 188, 191
Cycling test histories, electrode ... 210

D

Deformation rate 119
Demonstration cell test 212
Densities, mix 89
Design constraints for battery
  container material 223
Dewetting 188
p-Dichlorobenzene 175
Dicyclopentadiene 65
Differential scanning calorimetry . 16
Differential thermal analysis 177
Dimer, methylcyclopentadiene ... 65
Dipentene 65
Discharge
  curves 221
  reaction 220
  records, charge- 207
Dissolution rate 36
Double pond 62
Drum, sulfur-coating 39, 45
Duriez stability 141, 143
Duriez test 152
Dust control, sulfur coatings for .. 64
Dust formation, sulfur 50

E

Elastomer, carbon disulfide–
  unsoluble sulfur 21
Electrochemical reactions, anodic . 196
Electrochemical reduction of sulfur 196
Electrode
  carbon working 194
  copper sulfide 209, 210
  iron sulfide 211

Electrode (*continued*)
  lithium 187, 188
  metal sulfide 211
  sulfur 188, 197, 204
  surface, sulfur formation on the 197
  transition metal sulfide 203
Electrolyte
  failure mechanisms 218
  LiCl–KCl 187, 204
  sodium aluminate solid 218
  –sulfur batteries, sodium–solid . 217
  –sulfur compound, interactions .. 200
Energy storage 216
Engineering applications of sulfur-
  based materials, civil 154
Erosion control 64
Eutectic 196, 197
Evaluation criteria 116
Evolved gas analysis 123
Expansion, thermal 121
Experimental site 148, 149

F

Fatigue
  life analysis 125, 126
  resistance 91
  tests 143
Fertilizer 18, 33
Field tests 67, 96
Fillers 160
Filter 41
Flammability of plastics 182
Flexibility 91
Flexural modulus 179
Flexural strength 69, 70, 172
Fluid mixes, semi 89
Fly ash 164
Foam coatings, sulfur 167
Foamed sulfur 163, 169–171
  wallboard 83
Foaming process 81
Freeze-thaw cycles 122, 171
Fundamental transverse frequency 122

G

Gas analysis, evolved 123
Gas, hydrogen sulfide 123
Gel permeation chromatography .. 16
Gradation, aggregate 87, 91
Granulometries 146
Graphite cases for cells 207
Gravel asphalt concrete 115
Gravels, base layer 146
Gypsum wallboard, sulfur– 80

H

Handling trials, mix 96
Hardened mixes 90
High temperature cells 225
Hot aggregate–asphalt–sulfur
  mixes 88
Hveem stability 111, 112
Hydraulic applications 99

INDEX

Hydraulic sulfur spraying
   system 49, 52
Hydrogen sulfide
   concentration 158, 159
   evolution 124
   gas, toxicity 123
   levels 157
Hypersulfide 195

I

Illitic shale 162
Impermeability 93
Insulating value 165
Intergranular sulfur phase 22
Investigative cell test 209
Iron sulfide cells, lithium– 207
Iron sulfide electrode tests 211

J

Journal bearing data, comparative 182

K

Kaolinite 36

L

Laboratory
   evaluations 116
   penetration tests 64
   reaction tests 66
   test cell, sodium–sulfur 219
Leveling courses 97
LiCl–KCl electrolyte 187, 204
Li–KCl eutectic 196
Lightweight sulfur–gypsum
   wallboard 79
Limestone asphaltic concrete,
   crushed 115
Limonene polysulfides 11
Lithium
   batteries, secondary 203
   –copper sulfide cells 207
   couple, metallic 204
   electrode 187, 188
   –iron sulfide cells 207
   –metal sulfide cells 205, 206, 212
   –nickel sulfide cells 207
   polysulfides 204
   –sulfur cells 186
   –sulfur secondary batteries 186
Load leveling 216
Loss modulus 139, 140
Lubricants, solid 181

M

Marshall stability
   90, 108, 109, 116–118, 142, 152
Mastic mixes, sulfur-extended 99
Material(s)
   battery container 223
   for SAS mixtures 105
   variables 117

Metal sulfide
   active materials 208
   cells, lithium– 205, 206, 212
   couple, bivalent 204
   electrodes, transition 203, 211
Methylcyclopentadiene dimer 65
Microcrystalline wax 37
Microorganisms, sulfur-oxidizing .. 168
Mineral additives 163
Mist formation, sulfur 50
Mixes(es) (*see also* Mixtures)
   air voids content for 94
   asphalt paving 85
   densities 89
   designs 161
   handling trials 96
   hardened 90
   high bearing capacity 98
   hot aggregate–asphalt–sulfur .. 85, 88
   mastic 99
   permeability for 94
   porous 98
   preparation 90
   sand–asphalt–sulfur
      86, 97, 103, 105, 121
   stability 92
   stiff and semi-fluid 89
Mixing temperatures 118
Mixing time 116
Mixture(s) (*see also* Mixes)
   preparation of 148
   ratio 121
   sulfur–additive 198, 199
   sulfur–sand 78
   sulfur–selenium 199
Modified sulfur materials 15
Module characteristics, battery ... 214
Molding grades 179
Molding resins 182
Molecular weights 16
Molybdenum disulfide 181
Molybdenum sulfide 204
Monoclinic sulfur 4, 6
Mortar sands 79
Mortar, sulfur 76, 79
Myrcene polysulfides 11

N

Nickel sulfide cell, lithium– 207
Nitrogen fertilizers, controlled-
   release 18, 29
Nomenclature 20
Nozzles, pneumatic spray 41, 42, 44, 52

O

Orthorhombic sulfur 4

P

Panels, sulfur bonded prefabricated brick 76
Particle size distribution 79

Pavement(s)
  patching 98
  structure bases and surfacing .. 97
  sulfur–asphalt 102
  test sections 125
Paving mixtures, sulfur/asphalt 85, 141
Paving trials 96
Penetrability 134, 135
Penetration tests 65
Permeability 94
Pilot plant 33, 38
Plasticization of sulfur 1
Plastics 182
Pneumatic sulfur spraying system 40, 41
Pollution abatement, air 51
Polyethylene tetrasulfide polymers 2
Polymer(s)
  CTLA 65
  polyethylene tetrasulfide 2
  properties 176
Polymeric polysulfides 2
Polymeric sulfur 3, 4
Polyphenylene sulfide
  (PPS) 174, 176–179, 181–183
Polysulfide 189, 195
  alloocimene 11
  fractions 11
  limonene 11
  lithium 204
  from LP-33/sulfur reaction 12
  melt, sulfur 222
  myrcene 11
  polymeric 2
  sulfur rank in 15
  sulfur–sodium 220
  styrene 12
Polytetrafluoroethylene (PTFE) 181, 183
Ponds 60, 61
Porous mixes 98
Potassium vapor pressure 187
Potential scan rate 191
Prefabricated brick panels,
  sulfur bonded 76
Pressure-atomized sulfur coatings . 24
Pressure, hydraulic 51
Process temperatures 45, 52
Process variables 116
Pump, sulfur 49

R

Raw materials 35
Reaction tests, laboratory 66
Reduction peaks 194
Reduction, sulfur 189, 196
Release fertilizer, slow 33
Resins, molding 182
Reversion rate 6
Rupture modulus 159, 162
Ryton polyphenylene sulfide 177

S

Safety hazards 149
Sample preparation 116

Samplings, initial 124
Sand(s)
  I and II (Asphalt I),
    sulfur–asphalt 113
  asphalt–sulfur
    mixes 86, 97, 103, 105, 121
  base layer 145
  mixtures, sulfur– 78, 80
  mortar 79
Scan, anodic 194
Scan rate, potential 191
Sealants 37
  sulfur-coated urea with 40, 49
  sulfur-coated urea without 46, 53
Secondary lithium batteries 203
Segregation, aggregate and sulfur . 90
Selenium mixtures, sulfur– 199
Self-discharge mechanism 211
Setra formula mixtures 141
Shale, illitic 162
Shear rate 137
Shell Canada Limited 102
Shock failure 171
Shore D hardness 4
Sodium
  aluminate solid electrolytes 218
  anode assembly 224
  polysulfide, sulfur 220
  –solid electrolyte–sulfur batteries 217
  sulfide 174
  –sulfur
    batteries 216, 217, 221, 223
    cell 219, 225
    system, phase diagram 220
  sulfur reduction by 219
Softening point 134, 135
Soils 160
Solid lubricants 181
Solubility 132
Specific
  energy 216
  heat 122
  weight 134, 135, 141, 142
Specimen preparation 90, 168
Spectrophotometry 189
Splitting tensile strength 118, 119
Spray
  formulation 64, 69
  header system, sulfur 42
  nozzles, pneumatic 41
  plot data 68, 70
  system, hydraulic sulfur 49, 52
Stability
  Hveem 111, 112
  Marshall 90, 109, 116–118
  mix 92
  sulfur/asphalt 150
Stabilizing agent 64
Steel, coatings on 183
Stiff mixes 89
Strength
  compressive 80, 158, 162, 171
  effect of hydrogen sulfide on ... 157

INDEX

Strength *(continued)*
  effect of temperature on 159
  flexural 69, 70, 172
Stress distribution 156
Structural materials 55
Structure bases, pavement 97
Styrene
  polysulfides 12
Sulfur 36
  –additive mixtures 162, 198, 199
  allotropes 3, 21
  amorphous 21
  –arsenic mixtures 199
  –asphalt
    binders
      application of 150
      in high performance mixtures 141
      in low quality mixtures ... 144
      mixing of 150
      preparation 130, 148
      properties of
        133–135, 138, 139, 141–144, 150
    for road construction 130
    formulation 141
    mixes, sand– 97, 113
    pavements 85, 102, 141
  –based materials 154
  batteries, sodium– (*see* Sodium–
    sulfur) 216 217, 221, 223
  –bonded prefabricated brick
    panels 76
  –bound composites 160
  –brick interface 78
  building 57
  cells, lithium– 186
  chemical and physical properties
    of 154
  –coated block specimen 169, 172
  –coated urea (SCU) 19
    in the future 54
    nitrogen release from 29
    pilot plant 19, 33
    quality 43
    with sealant 40, 49
    without sealant 46, 53
    texture of 18
  coating(s)
    air-atomized 20
    applications of 167
    defects in 26
    drum 39, 45
    for dust and water erosion
      control 64
    pressure-atomized 24
    properties of 19
  combined 15
  in construction materials 75
  containing composites 165
  content 92, 108, 110, 112
  creep behavior of 161
  –cretes 160
  crystallization 5, 8, 9
  dark 77

Sulfur *(continued)*
  dust 50
  elastomer, carbon disulfide–
    insolubility 21
  electrode 188, 197, 204
  on the electrode surface 197
  elemental 157
  in the eutectic 197
  -extended mastic mixes 99
  flowers 15
    foam coatings of 167
    production of 163
    specimens 169–171
  free 15
  freeze–thaw failure in 171
  –gypsum wallboard, lightweight 79
  in hot mixes 85
  insulating value of 165
  in the LiCl–KCl eutectic 196
  liquid 104
  loss 197, 198
  materials, modified 6, 9
  materials, thermal shock failure in 171
  with mineral additives 163
  mist 50
  mixes, hot aggregate asphalt– .. 88
  mixes, of sand–asphalt– 86, 103
  monoclinic 4
  mortar 76, 79
  nozzles 42, 44
  orthorhombic 4, 6
  -oxidizing microorganisms 168
  phase, intergranular 22
  plasticization of 1
  polymeric 3, 4
  –polysulfide melt 222
  production, Canadian 155
  pump 49
  rank in polysulfides 15
  reaction, LP-33- 12
  reduction 189, 219
  –sand mixtures 78, 80
  secondary batteries, lithium– ... 186
  segregation, aggregate and 90
  –sodium polysulfide 220
  solubility of 132
  specimen, fractured 71
  specimen, quenched 156
  spray header system 42
  spraying systems 40, 49
  as a stabilizing agent 64
  in structural materials 55
  wallboard 82, 83
Supersulfide 189, 195
Surface
  bonding 61
  layers 147
  overlay, bleeding 98
Surfacing, pavement structure 97

T

Tailings, copper 64

Temperature
  curve, viscosity– 104
  cyclic voltammograms as a function of 191
  distribution in a quenched sulfur specimen 156
  effects of 160
    on flexural modulus 179
    on splitting tensile strength .. 119
    on strength 159
  process 45, 52
Tensile strength 4, 118, 119
Test methods 116
Thaw cycles, freeze– 122, 171
Theoretical analysis, preliminary .. 125
Thermal
  blanket applications 165
  conductivity 120
  expansion 121
  properties 120, 121
  shock failure 171
Thermopave 102
Thiokol 2
Thiols 16
Toxicity of hydrogen sulfide gas .. 123
Toxicity of the spray formulation .. 64
Traffic loadings 116

Transition metal sulfide electrodes 203
Transverse frequency, fundamental 122

U

Urea 35
  sulfur-coated
    (*see* Sulfur-coated urea)

V

Vapor pressure, potassium 187
Vapor pressure of sulfur–additive mixtures 198
Viscoelastic properties 139
Viscosity 104, 133, 137
Voids content, air 94
Voltammetry, cyclic 188, 191

W

Wallboard
  foamed sulfur 82, 83
  lightweight sufur–gypsum 80
  production process 83
Water erosion control 64
Water on sulfur with mineral additives, effects of 163
Wax 28, 37

The text of this book is set in 10 point Caledonia with two points of leading. The chapter numerals are set in 30 point Garamond; the chapter titles are set in 18 point Garamond Bold.

The book is printed offset on Danforth 550 Machine Blue White text, 50-pound. The cover is Joanna Book Binding blue linen.

Jacket design by Linda McKnight.
Editing and production by Virginia Orr.

The book was composed by the Mills-Frizell-Evans Co., Baltimore, Md., printed and bound by The Maple Press Co., York, Pa.